Volcanoes

Volcanoes

Crucibles of Change

RICHARD V. FISHER

GRANT HEIKEN

AND

JEFFREY B. HULEN

ILLUSTRATED BY JEFF
AND RENATE HULEN

Princeton University Press

Princeton, New Jersey

Library of Congress Cataloging-in-Publication Data
Fisher, Richard V. (Richard Virgil), 1928–
Volcanoes, crucibles of change / by Richard V. Fisher, Grant Heiken,
and Jeffrey B. Hulen.
Includes bibliographical references (p.   –   ) and index.
ISBN 0-691-01213-X (cloth : alk. paper)
1. Volcanoes.   I. Heiken, Grant.   II. Hulen, Jeffrey B. (Jeffrey
Brooke), 1946–.   III. Title.
QE522.F57   1997
551.21—dc21   96-49516

The epigraph is from the poem "The Homing Call of Earth,"
by Kofi Anyidoho, in *Earthchild* (Accra: Woeli
Publication Services, 1985), pp. 43–44.

This book has been composed in Times Roman and Helvetica
Designed by Jan Lilly

Princeton University Press books are printed on
acid-free paper and meet the guidelines for
permanence and durability of the Committee on
Production Guidelines for Book Longevity of
the Council on Library Resources

Printed in the United States of America
by Princeton Academic Press

1   2   3   4   5   6   7   8   9   10

*We dedicate this book to our colleagues
who have died as a result of volcanic eruptions
while adding to our knowledge of volcanoes*

CONTENTS

Our Earth survives recurring furies
  of her stomach pains and quakes
From the bleeding anger of her wounds
  volcanic ash becomes the hope
  that gives rebirth to abundance of seedtimes.

Kofi Anyidoho, "The Homing Call of Earth"

Frontispiece. Majestic Mount Shasta, California, the largest of the Cascade volcanoes of the Pacific Northwest, symbolizes the beneficial coexistence of man and nature. Active volcanoes replenish the atmosphere, oceans, and soils of our planet. Volcanoes can inspire spiritual awe and humility, and they amplify the realization that we must live in balance with and in deference to nature, of which we are a part. Although the volcano is currently dormant, it has had a violent past, and we must remain apprised of its future. (Photo: Richard V. Fisher)

Volcanoes are crucibles of change and are held in fascination mainly for their awesome powers of destruction. In the span of human life, they can change a landscape from jungle to desert, degrade the global climate, induce great floods, and even bury entire cities. Yet volcanoes are ultimately benevolent. The oceans in which life began, and the lakes, rivers, and groundwaters that renew and sustain life, are all condensed volcanic steam that was produced during countless eruptions over billions of years. The very air we breathe is a store of volcanic vapors. Wondrously fertile volcanic soils yield sustenance for millions in tropical and temperate regions. Volcanic geothermal systems that breach the surface as soothing hot springs are also clean, safe, and renewable sources of electrical energy. The tallest volcanoes wring moisture from passing clouds, creating glaciers as well as tumbling rivers that endow us with hydroelectric power. Primitive dwellings are carved in soft but strong tuffs, and these and other volcanic rocks are among the most commonly quarried construction materials. Precious stones and metals have long lured adventurers to volcanic terrains. The value of these tangible goods is greatly enhanced by the scenic beauty and commanding spiritual presence of the volcanoes themselves.

The ancients saw gods within all natural phenomena—fire and rain, sun and moon, all manner of life, earthquakes, and especially volcanoes—imbuing them with human traits. The volcano gods, unless adored and appeased, would torment or even casually kill the people who inhabited their mountainous domains. Knowledge gained from scientific research is the intelligent way of mitigating volcanic hazards, but there are still those people who make offerings to volcanoes in the hope of preventing their deadly eruptions.

What actually triggers these eruptions? When and where will the next one occur? How large and how destructive will it be? As the world's population approaches six billion (and possibly twelve billion by 2035), answering these questions becomes increasingly critical because many of these people inevitably will settle on and around dangerous volcanoes. Although volcanologists have greatly improved the basic knowledge of eruption mechanisms, their ability to forecast these events is still far from perfect. They are still searching for new ways to predict volcanic eruptions and their potentially disastrous effects with ever greater accuracy.

Alleviating these effects, however, is not entirely a scientific problem; equally important are social and political forces. Even when warned of imminent eruption, people who live on or near the volcano are often reluctant to leave. The entreaties of scientists, frequently discounted, are far outweighed by the urge to remain and protect homes and livelihoods. It is far easier and more comforting to heed a venerated elder, who might proclaim, "I have lived here eighty-five years, and in all that time this volcano has never erupted." Yet eruptions may be separated by extended quiet intervals and still have the potential to scour the volcano and its surroundings of all living things. Accordingly, volcanologists have a mandate not only to better understand volcanic processes, but to convey the essence of this new knowledge to the people who may suffer from the effects of volcanoes. Only in this way can the certain ravages of future eruptions be kept to a minimum.

We begin this book with an account of the May 1980 eruption of Mount St. Helens in Washington State, an event that exemplifies the enormous complexities of volcanic risk mitigation managed by government officials and volcanologists working under duress with a public that was variously curious, concerned, frightened, and confrontational. Subsequent chapters describe the rich variety of volcanoes, the reasons for their eruption, the different kinds of volcanic hazards, and the impact of volcanism on all aspects of human life. We conclude by mapping a course for prospective volcano travelers.

Much of what we have written comes from various aspects of our own work or from excursions to the field areas and laboratories of other researchers. We liberally use the findings and experience of other workers in volcanology and thank them for their contributions to the understanding of volcanism and its effect upon human beings. In particular, we are indebted to Ray E. Wilcox, who is now retired from the U.S. Geological Survey after many years of scholarly volcanological research. Dr. Wilcox was one of the first to analyze the range of effects of volcanic eruptions on public health and safety, utilities, transportation, communication, and agriculture in his 1959 U.S. Geological Survey Bulletin *Some Effects of Recent Volcanic Ash Falls with Especial Reference to Alaska.*

Our book is not a comprehensive look at volcanology. Rather, we emphasize the effects of volcanic eruptions on society, viewing volcanoes as among nature's most important crucibles of change—initially destructive, but in compensation, the source of enduring material and spiritual wealth.

Colleagues who gladly shared their experiences and photographs are José Viramonte, Vince Neal, Rick Wunderman, Dick Fiske, Jim Luhr, Tom Simkin, Barry Voight, and Floyd McCoy. Chris Newhall, of the U.S. Geological Survey, gave us inspirational advice, especially about Mount Pinatubo. Giovanni Orsi and Lucia Civetta introduced us to the remarkable geology of Naples, where people and volcanoes mix every day; we owe them a considerable debt for their help and friendship. GH thanks Giday WoldeGabriel and Tim White for involving him in the Middle Awash Project, in which volcanism and the early history of man are inextricably linked. GH also thanks Pat Browne of New Zealand's Geothermal Institute for his considerable insights on geothermal energy and historical eruptions in New Zealand. The staff at the U.S. National Archives, Still Photo Branch, were extremely helpful in locating old eruption photos. JBH thanks Joe Moore and Mike Wright for helpful discussions on volcanic geothermal systems and mineral deposits as well as geothermal power production. Photographs credited to Harry Glicken were given by him to RVF; Harry died in a volcanic eruption at Mount Unzen, Japan, in 1991. Peter Frenzen, Monument Scientist for the Mount St. Helens National Monument, freely offered information about the monument. Beverly Fisher located the African poem (in the front of the book) that so appropriately sums up the content of this book.

Drs. Richard Waitt and Norman Banks, members of the volcano crisis team of the U.S. Geological Survey during the eruption of Mount St. Helens in 1980, helped to improve chapter 1, in which we discuss that eruption. We thank Dr. Michael Lyver, research psychologist at Bond University, Australia, for his impressions of the Mount Unzen eruption in early June 1991, which are included in chapter 5. Dr. Robert Tilling, of the U.S. Geological Survey, gave especially helpful suggestions for improving chapter 15 on the mitigation of volcanic eruptions. Dr. Donald Peterson, also of the U.S. Geological Survey, kindly loaned published and unpublished materials on hazards mitigation. John Bares of the Robotics Institute, Carnegie Mellon University, sent us information on the Dante Project.

The book could not have been written without the early guidance of our teachers, the years of communication with and fellowship of our colleagues, and the countless questions that were asked by students who made us think and taught us reciprocal lessons that widened our horizons.

We thank Peter Francis for his thorough and helpful reviews of this book.

PART I

Volcanoes and Eruptions

Politicians and Volcanoes

Frontispiece. A herd of elk crosses forest devastated by the May 18, 1980, eruption of Mount St. Helens, Washington. No matter how desolate an area may be after a volcanic eruption, life is quickly reestablished and can flourish after a few decades. (Photo: Richard V. Fisher)

Fig. 1-1. Washing clothes along the Rivière Roxelane—a typical day in the city of St. Pierre, Martinique, before the eruption of Mount Pelée in 1902. The sturdy stone houses were characteristic of this port city, which was then known as the Paris of the Caribbean. (Photo: Heilprin, 1908; Geographical Society of Philadelphia)

It was a coincidence of geography. On May 8, 1902, the beautiful city of St. Pierre, Martinique, known as the Paris of the Caribbean, happened to be in the killing path of a searingly hot volcanic hurricane that swept down the flank of Mount Pelée and killed 29,000 people. And it was a circumstance of timing that the city was in the process of election campaigning. It was an atmosphere doubly charged with election politics and a volcano that had been acting up for a couple of months. People were ambivalent about leaving: on the one hand, they wanted to support their candidate, but on the other, the explosions, rumblings, and sulfurous smell of the volcano frightened many, some of whom chose to leave the area. And yet, there were some curious people who came to St. Pierre from other towns and villages to watch the eruption. In 1902 volcanologists had a very rudimentary knowledge about volcanic processes and did not know that volcanic hurricanes (*pyroclastic flows*: *pyro* = "fire"; *clastic* = "broken") existed. Hence this eruption made Mount Pelée the greatest killer volcano of the twentieth century. If the precatastrophic volcanic activity then shown by Mount Pelée were to occur today, modern volcanologists, having gained ninety years of cumulative experience since the 1902 eruption, would very likely urge immediate evacuation of the city regardless of other circumstances.

Gaston Landes, a professor at the high school in St. Pierre in 1902, said that ash and the sulfur smell from Mount Pelée would not cause great damage, and if the volcano erupted lava flows, they could not reach the city because too many ridges and valleys stood in the way. He was right about the lava flows, but he did not know about hot pyroclastic flows. He gave assurances that little damage would occur even with a big eruption. But on May 8 at 8:03 A.M. the most explosive eruption began, and within five minutes the pyroclastic flow swept through St. Pierre, blowing away, burying, and burning buildings and people. Those not killed outright by physical force died by inhaling ash, which clogged their nasal passages and trachea, and searing gases, which scorched their lungs (figs. 1-1 and 1-2).

The eruption of Mount Pelée drew the world's attention to the existence of pyroclastic flows and started a line of research that is still continuing (see chap. 5). Since 1902, volcanologists have gathered considerable information about the behavior of volcanoes. They now have instruments to detect the small swellings of a volcano that indicate molten rock (*magma*) is rising into

Fig. 1-2. The eruption of Mount Pelée on May 8, 1902, devastated much of St. Pierre, leaving piles of rubble and half-standing walls. Sturdy stone walls were knocked down by powerful fast-moving clouds of hot dusty gas, known as pyroclastic surges or "volcanic hurricanes." Note that the volcanic hurricane left no discernible volcanic deposit on the rubble that covered Rue Victor Hugo. Nearly all of St. Pierre's residents died from the blast, burns, or asphyxiation. (Photo: Heilprin on May 14, 1902. Published in 1908 by the Geographical Society of Philadelphia)

the edifice of the volcano. Seismologists, using seismographs, can find the origin of an earthquake (the *focus*) below a volcano and pinpoint the depth at which magma is stirring. Gases that originate from magma and reach the surface can be analyzed. One of these gases is sulfur dioxide, which indicates that live magma is moving upward through the volcano and may erupt. The presence of older rock fragments within the exploded debris suggests that the explosions are caused by groundwater that flashes to steam from the heat of the underlying magma and blasts apart preexisting solid rock. The presence of newly solidified pieces of magma suggests that magma could be approaching the surface.

Knowledge of volcanic behavior gained during the twentieth century was used by volcanologists at Mount St. Helens in 1980 to warn of impending danger, but as with Mount Pelée, volcanologists still did not have enough information to predict with certainty how Mount St. Helens was going to behave.

A Modern Tragedy:
Mount St. Helens

On March 25, 1980, a group of geology students and three professors (including one of the authors—RVF) from a class in volcanology at the University of California at Santa Barbara visited the Hawaii Volcanological Observatory. We went to the wrong place to witness the beginning of an eruption, for on March 27, Mount St. Helens would erupt. But we did learn from an excited seismologist at the observatory that an earthquake of magnitude 4.1 had occurred beneath Mount St. Helens in southern Washington on March 20, ending the volcano's 123-year rest. Other earthquakes at Mount St. Helens before March 27 had caused cracks to appear in its summit glaciers. Had we gone to the chain of volcanoes that forms the backbone of the Cascade Mountains instead of to Hawaii, we would have witnessed the modest beginnings of one of the most highly publicized and intensely studied eruptions of the twentieth century. Then, on May 18, 1980, nearly two months after the first stirrings of Mount St. Helens, the volcano turned violent.

An Eruption Carnival Turned Tragic

Mount St. Helens was center stage and people came for the show. During the interval prior to its climactic eruption—from March 20 to May 17, 1980—the volcano entertained the public. Mount St. Helens' antics and its potential for danger drew people from everywhere. Entrepreneurs sold T-shirts, cups, posters, and bric-a-brac labeled with humorous slogans and cartoons (fig. 1-3). The main evacuation artery north of Mount St. Helens, Washington State Highway 504, was jammed with sightseers, there to witness the first eruption in the lower forty-eight states since Mount Lassen erupted in 1914. Some people said they came to see the "big one." Cups of coffee and snacks were sold by the thousands. Roadblocks challenged some of the spectators, who then broke them down or snuck around them. State officials set up a "red zone" around Mount St. Helens to keep people from possible harm should there be an eruption (fig. 1-4); however there were not enough personnel to monitor the roadblocks twenty-four hours a day. The activity of the volcano excited people, and they wanted to see it. But at 8:32 A.M., May 18, 1980, the carnival turned tragic with a violent eruption. Many people, some on official business, died within the red zone. Volcano watchers, as well as campers thought to be in safe zones and out of sight of the volcano, were also caught in the doomed area.

On the quiet, sunny morning of May 18, Charles McNerney, John Smart, and several other people had found their way to an overlook to watch Mount St. Helens, just in case it erupted. They had a good view of the volcano from

Fig. 1-3. Before the devastating blast of Mount St. Helens on May 18, 1980, the road to the mountain was clogged by thousands of sightseers, creating a carnival atmosphere along Highway 504. Hundreds of people attempted to evade roadblocks to witness minor steam eruptions that preceded the eruption of May 18. No one had an inkling of the catastrophe that lay ahead. (Photo: Courtesy of Harry Glicken)

a cleared area along the North Toutle River near Castle Lake, 13 kilometers (see appendix 2) west of the volcano. At 8:32 A.M., the volcano watchers got their wish. It was much deadlier than they could have dreamed, for the north side of the volcano collapsed. Then came the blast. A black cloud came directly from the summit and within seconds climbed over a ridge toward them. A warm wind began blowing ahead of the cloud and increased until trees bent over and branches broke; the approaching blast must have pushed air in front of it. The events of the first two minutes after the eruption were warning enough for McNerney and Smart to leave. The other people may have decided to leave a minute or two later, but it would have been too late because even McNerney and Smart barely escaped with their lives. They drove as fast as they could (up to 125 kilometers per hour on straight stretches) down a dirt road toward Highway 504 but the cloud was gaining on them. They felt the radiant heat of the cloud carried by the wind that blew into the car through the open sunroof. It felt like the car's heater was on. After reaching the paved Highway 154, they accelerated to 140 kilometers per hour on some stretches

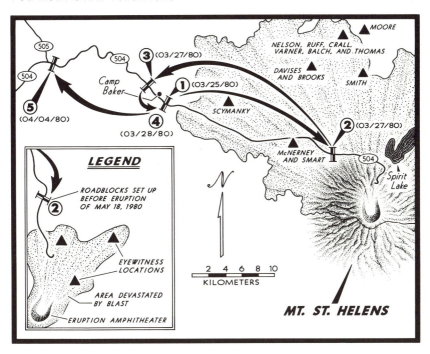

Fig. 1-4. The ambivalence of the authorities about the dangers of the impending eruption of Mount St. Helens is shown by the continual movement of the roadblocks that were set up to keep people away from the dangers of the volcano. The placement of the road-blocks turned out to be partly politically motivated in an attempt to protect citizens from themselves. An explanation of the chronology, which is indicated by numbers, is given in the text.

and finally began to outdistance the cloud. When they lost sight of it, they stopped in relief, hoping that it, too, had stopped; but it soon reappeared, moving at about 72 kilometers per hour. The base of the cloud looked to them like an avalanche of black chalk dust. First, one part of the cloud shot out in front, then another, then another, like waves lapping onto a beach. What they were seeing was the turbulent front of a deadly pyroclastic flow. After pulling back onto the highway, they finally outran the cloud, at an average speed of 105 kilometers per hour.

Early that morning, James Scymanky and three others were cutting wood in the valley of Hoffstadt Creek, 20 kilometers away from, and completely out of sight of, Mount St. Helens. Showing little concern about the volcano, Scymanky and two other men were felling small trees with chainsaws when a fourth man shouted that he saw a cloud appear over the ridge. Within about ten seconds came a thunderous crashing, crunching, and grinding sound from the east as trees were felled by of the blast. The men were soon engulfed by the blast cloud. The force knocked Scymanky down, and he felt a searing, painful heat on his back that lasted about two minutes. When visibility returned, the forest of small trees lay upon the ground throughout the entire valley, which was covered with 25 centimeters of gray ash. The men's clothing remained intact, but their bodies had been extensively burned. Scymanky's lungs were not fatally seared, and he somehow survived his burns, but the others subsequently died from theirs.

The explosion from Mount St. Helens fanned outward on the ground from the volcano in the pattern of a half circle. It had occurred as a blast that moved faster and more forcefully northward than eastward or westward. From eyewitness accounts, estimates of the cloud's northerly velocity ranged from 496 to 510 kilometers per hour for a distance of 13 kilometers. Four parties of campers had settled near the northern boundary of destruction. They were not volcano watchers and had not camped where they could see the volcano. They were also separated from it by one or two 300-meter-high ridges. Three of these people, Dale and Leslie Davis and Albert Brooks, had stalled their pickup truck while trying to escape and the blast cloud overran them. It looked to them like a "boiling mass of rock" that picked up trees and threw them into the air. Darkness descended as the cloud surrounded them, and it became very hot in the truck cab. Chunks of rock and wood pounded the truck for a short time but, except for a small vent on the right side of the truck, the windows did not break. One of the trio received burns on both legs, just above the ankles, as ash entered the truck through the broken vent window. A brief period of light appeared a few minutes after the blast first descended upon them, but then ash fell again so densely that they could not see the truck's hood. Nearly a half hour later, at 9 A.M., they decided to abandon the vehicle and walk, even though the darkness required that they use a flashlight. By 10:20 A.M. murky ash still kept visibility at about 10 meters. It wasn't until later that day, after the sky had cleared, that they were rescued by helicopter.

Later examination of the truck showed that heat had deformed its plastic grill and had melted a styrofoam cooler in the truck bed. The vehicle's right side, which had faced the volcano, had been extensively damaged—the blast had stripped its chrome trim and had sandblasted the paint.

In Miners Creek, which was over the ridge from the pickup truck, Edward Smith and his companions had tipped an external-frame tent on its side to dry out. A little after 8:32 A.M., several gusts of wind suddenly blew the tent over and were immediately followed by what sounded like three distant rifle shots. Ten to fifteen seconds later, a black cloud shot overhead, and rocks as big as golf balls began to fall. The sky became dark, and ash fell so heavily that visibility was no more than about a half a meter, even by the light of a flashlight. Although thousands of trees fell, Smith and his friends did not hear them, nor did they feel a blast or wind at that time. The first material to fall on them was cold, but shortly thereafter, the air became uncomfortably hot, as if a forest fire were nearby, but the heat did not last very long.

Two other groups were camped along the north bank of the Green River about 22 kilometers north of the volcano and across two 300-meter-high ridges, certainly a safe distance from Mount St. Helens, or so everybody thought, even the volcanologists. One of the groups included Bruce Nelson, Sue Ruff, Terry Crall, and Karen Varner. Nelson and Ruff told of a wind that preceded the blast cloud by ten to fifteen seconds, strong enough to blow flames horizontally outward from the campfire, but not strong enough to impede a person's movement or to blow down trees. The campers heard no loud noises prior to the cloud's arrival, but when it did arrive, daylight turned to darkness and all of the trees seemed to topple over simultaneously. In the confusion, Nelson and Ruff accidentally fell into a hole left by the root ball of a blown-over tree and were protected from other falling trees. They could hear one another while in the hole, but it was too dark to see. After they had been in the hole about ten seconds, the temperature increased enough to singe their hair, which they could hear sizzle. (Hair singes at 120°C.) Bruce Nelson, a baker familiar with ovens, estimated the temperature to be about 300°F. The heat was intense enough to boil pitch from the trees and still cause minor burns several minutes later. After a few minutes of darkness, the sky suddenly cleared for a few minutes and then a dense fall of ash began. After Nelson and Ruff had dug themselves out of the debris, they called for Crall and Varner but received no answer—they had been killed by falling trees. Close by were two other campers, Dan Balch and Bryan Thomas. A tree fell on Bryan Thomas and broke his hip, but he was too heavy to carry, so Nelson, Ruff, and Balch built him a lean-to for shelter and set out on foot. Balch, who was barefoot, could not walk through the hot ash and was also left behind. Nelson and Ruff continued and during their trek encountered Grant Christensen, whose pickup truck would not start. After about three hours, the three sighted a helicopter, which they signaled by using their clothes to stir up dust. The three men, together with Balch and Thomas, were rescued that day by helicopter.

Mike and Lu Moore, from the town of Castle Rock, were camped near the Green River at a point 20 kilometers from the volcano with their daughters, four-year-old Bonnie Lu and three-month-old Terra. They had parked and then hiked a 4-kilometer-long trail to spend Saturday night at a campsite intending to return to Castle Rock on Sunday afternoon. While making breakfast the next morning, Lu Moore noticed a rumbling noise and reported that "it felt like there was an earthquake inside you." The family quickly loaded their backpacks and moved to a shelter for hunters. According to Mike Moore, "You could physically see the cloud of ash moving toward us. It was the blackest cloud I had ever seen" (*Daily News*, Volcano, p. 61). Thunder and lightning accompanied the cloud, which passed over them, and it turned very dark as ash began to fall. So much ash rained down that the Moores had to breathe through stockings to filter it out. Within an hour or two, when the darkness had lifted, they wanted to return to their car but were blocked by fallen trees. They spent Sunday night in their tent and were rescued Monday morning by a helicopter. The Moore family had been just outside of the devastation boundary. Before the blast cloud reached them, it had risen from the ground and lofted over their heads, so rather than being overwhelmed by the hurricane-like cloud moving laterally across the ground, they were showered by ash.

Venus Dergan and Roald Reitan, who were camped alongside the Toutle River, woke up as the river began to rampage. As the water rose and they began throwing their belongings into the car, Reitan saw a huge logjam that was being held back by part of a railroad bridge. The water rose swiftly and they climbed to the top of their car, but the rising water swept the vehicle down a steep embankment and they had to jump off. Dergan was then sucked under, and when Reitan located her, he could only see her nose between two logs. Reitan reached her by pulling himself along one of the logs. He then grabbed her by the hair and pulled her to the surface. They somehow managed to reach the edge of the flood and ran in knee-deep muddy water to reach a road. Fearing that the water would soon wash out the road, they climbed up a steep hillside and escaped.

In the final count at Mount St. Helens, thirty-five people were confirmed dead and twenty-two more were never found. One hundred and thirteen people were rescued by helicopters on the day of the eruption, and within the first few days, rescuers had found a total of one hundred and twenty-eight survivors. Fifty-three people were found within 1.6 kilometers outside of the blast boundary. Forty-seven were rescued from the south side of the volcano, where there was neither a blast surge nor a heavy ash fall. Eleven survivors were just inside the blast-zone boundary. Of these, six were unharmed, two inside a vehicle had second-degree burns, two endured third-degree burns, and one suffered a fracture.

Autopsies of seventeen victims showed that they had died from asphyxiation by inhaled ash. Two victims died from a tree that fell on their tent. Another, parked in his automobile 16 kilometers from the volcano, died when he was hit by a rock that flew through the car window. The skin of many of the dead was mummified with dark discoloration, and exposed muscles were cooked rather than desiccated. Hair was singed on victims found as far as 18 kilometers from the volcano, and burns accounted for three deaths that occurred 15 kilometers from the summit. Burning, singeing and mummification were caused by the heat of the blast cloud.

Temperatures of the blast cloud were estimated by the effects of the heat on vehicles, trees, and buried wood. The most revealing effects were on plastics composing the turn-signal lights, tail lights, and instrument panels of the cars and trucks. To determine melting temperatures, similar materials were obtained from vehicles of the same type and year of manufacture and were exposed to heat. Estimates ranged from 100° to 300°C, yet some people survived because the hottest part of the blast cloud had moved swiftly by. The blast cloud from Mount St. Helens was relatively cool compared with many pyroclastic flows, which can attain 850°C or more.

Another peril stemming from the eruption was the *lahars* (mudflows; see chap. 6) that originated in three different ways at Mount St. Helens:

- Water within the pores of rocks of the original volcano, which collapsed to become an avalanche, was squeezed from the pores of the avalanche by its own weight after it had stopped. The water that oozed out of the stationary avalanche deposit ran from its surface in rivulets that coalesced into bigger and bigger lahars that raced down the North Fork Toutle River.
- Some lahars were fed with water from snow melted by hot pyroclastic flows.
- Pyroclastic flows plunged into the streams and rivers and mixed with the water to form thick mud that moved downstream as lahars.

Lahars from these diverse origins mixed at tributary junctions to create increasingly larger lahars. As they mixed with stream water, they became more dilute and turned into floods that carried enormous loads of ash down the canyons draining the volcano and then into the tributaries of the Columbia River. The floods took as many as six lives and caused millions of dollars in property damage as they swept down the north and south forks of the Toutle River (fig. 1-5). And so much ash was transported to the Columbia River that it became too shallow for ships sailing from the Pacific Ocean to the Port of Portland. Ship traffic was completely stopped for a week and then restricted for three months while the ashy muck was dredged from the river. The flooding also damaged port facilities along the Columbia River at Longview and Kelso, Washington, and automobile, truck, and railroad traffic between Seattle and Portland was interrupted by flooding and damage to bridges.

Far from the Volcano

The eruption's impact was not as immediately dramatic far from the volcano as it was in the blast zone, but it caused serious social and economic disruption across eastern Washington, Idaho, and Montana—a populated area in contrast with the wilderness around Mount St. Helens. In eastern Washington, 152 million cubic meters of volcanic ash fell like snow upon the cities and towns and millions of acres of agricultural land. Almost 50 percent of Washington State felt the greatest impact from the various effects of the volcano. More than 4 centimeters of ash fell on Ritzville, 275 kilometers from the volcano, but only 1.5 centimeters fell on Yakima, 135 kilometers northeast of

Fig. 1-5. Debris flows along the Toutle River caused severe damage to homes and logging camps. This photograph shows a crushed house and an overturned mobile home. The debris flows were caused by drainage from the avalanche in the North Fork Toutle River and from pyroclastic flows that mixed with water in the South Fork. The debris raced down rivers draining Mount St. Helens after the May 18 eruption. (Photo: Courtesy of Harry Glicken)

Mount St. Helens. Low-level winds apparently deposited ash on Yakima, but the eruption cloud that penetrated high into the atmosphere rained down most heavily in the Ritzville area. Spokane, 410 kilometers away, received about 0.5 centimeter of the talcum-powder-like volcanic dust. Farther across the United States, the highly dispersed and fine-grained ash was not recognized by those unaware that dust from Mount St. Helens was in the air (fig. 1-6).

Although the layer of ash was thin, it was enough to disrupt power, communication, and transportation systems across eastern Washington. Airports and highways were closed for up to a week. Where rain had fallen, electrical transmission facilities discharged as a result of arcing across the power-line insulators that were coated with wet ash. Because of darkness and unexpectedly cold daytime temperatures, power usage by residents who were turning on lights and heaters taxed generating plants. Telephone services became overloaded by concerned people calling neighbors and friends.

It was soon discovered that shoveling volcanic ash was not like shoveling snow. The ash was hard to handle: as a dry powder, it was exceedingly difficult to contain on shovels, and as a wet mixture, it was much heavier than snow.

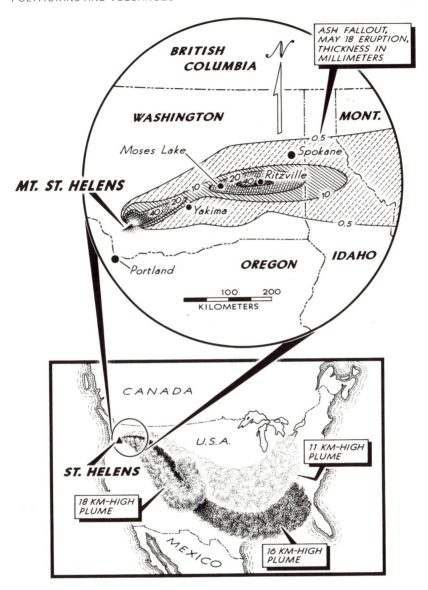

Fig. 1-6. Effects of the May 18 Mount St. Helens eruption were felt well beyond the Cascade Range of the Pacific Northwest. High-altitude winds carried volcanic ash across the United States and eventually around the world. Ash fallout ranged from thin dustings in Denver, Colorado, to 40 millimeters in Ritzville, Washington.

Because the ash did not melt, it had to be removed and places found to dump it, all of which added to the cost of disposal.

The ashfall from Mount St. Helens was relatively small in volume when compared with that from truly large eruptions, but the problems that modern society faces from even a relatively small ashfall are illustrated by the experience of Yakima (pop. 50,000). The ash cloud first reached the city at about 10:00 A.M., an hour and a half after the blast. Lightning and thunder accompanied the dark cloud. Cars stalled, motors and generators stopped running, and the city was silent and dark for sixteen hours. An estimated 546 million kilograms of ash fell on Yakima, with nearly 14.5 million blanketing airport runways. All highways and the Yakima airport were closed, and the city was isolated. The tiny, sharp glassy ash particles choked air filters and abraded bearings in wheels and motors. Indoors, dusting the ash off furniture left scratches that looked as though they had been caused by kitchen cleanser.

Yakima had no planned volcano emergency procedures or experience with ashfalls. Local and state officials tried to contact emergency services for information, but the offices were unprepared for ashfall problems. The Emergency Broadcast System, known to dedicated radio listeners throughout the United States, did not provide a warning because the Sunday-morning staff did not know how to operate the equipment. The air was dusty and murky, roads were closed, and there were no buses, planes, or trains available. Mail deliveries and refuse collection would not resume on Monday, nor would the schools open. The sewage treatment plant became clogged with ash and could not operate. Despite the threat of disease, raw sewage had to be discharged into the Yakima River. A local radio station that furnished twenty-four-hour news proved to be the most useful medium for disseminating information.

It quickly became evident that equipment and manpower were inadequate to clean up the city and that outside help was needed. The entire cleanup eventually required ten times more money than was needed for usual maintenance. The costs were also indirect. During the cleanup, Yakima had to suspend business activity. The central business district was closed for two and one half days while streets, alleys, sidewalks, parking lots, and roofs were swept and shoveled clean of ash. Any modern city downwind from active volcanoes needs to study the Yakima experience. Only 1.5 centimeters of ash fell upon the city, but the problems and costs were enormous, and these problems multiply geometrically with increasing ash thickness.

The Costs of the Eruption

The $2.7 billion in damages caused by the eruption of Mount St. Helens affected all of the taxpayers in the state of Washington, and in one way or another, in the entire nation. Short-term losses to Washington State cost as much as $970 million. Over half of the losses were timber resources. Cleanup costs were another $270 million, and $85 million was spent to restore damaged property (roads, bridges, and other property in the blast zone and flooded areas). Agricultural losses were estimated at $40 million. Twenty-seven bridges and 272 kilometers of roads had to be replaced. Sixty-one houses were totally destroyed, fifty-five were heavily damaged, and sixty-four were isolated by mudflows along the Toutle River. Two-thirds of the cleanup costs went to areas near the volcano and downstream to the Columbia River. Substantial funds went to dredging the Toutle, Cowlitz, and Columbia Rivers.

The Eruption Setting

Mount St. Helens straddles Cowlitz and Skamania Counties, is bordered on the north by Lewis County, and lies within the U.S. Forest Service's Gifford Pinchot National Forest of southern Washington. The Green River drains the area north of the volcano, the South Fork and North Fork of the Toutle River drain its western side, and the Lewis River—along which there are three dams that form Lake Merwin, Yale Lake, and Swift Reservoir—drains the southern side of the mountain. Their waters flow into the Columbia River and on to the Pacific Ocean.

Before the eruption, Mount St. Helens was one of the premier outdoor playgrounds in the state of Washington, being easily accessible to the major population centers of Portland and Seattle. The once graceful conical shape loomed high over Spirit Lake, whose shoreline was dotted with organization camps and resorts. The area was an outdoor paradise for fishing, camping, hiking, skiing, and hunting. The land north of the volcano harbored many small fishing lakes and was graced by old-growth Douglas fir, hemlock, and noble fir. Crusty Harry Truman, in his eighties, owned and ran a resort on the south side of the lake nearest the volcano. It was a land of timber, the largest active company in the area being the Weyerhaeuser Company, with headquarters at Longview, Washington.

In addition to land held by the Gifford Pinchot National Forest, Mount St. Helens and the surrounding land was owned by a variety of private and government concerns. The western volcano margin was owned principally by the Weyerhaeuser Company, its closest border about 5 kilometers away from the volcano. On the mountain and to the north, alternate sections of land were owned primarily by Weyerhaeuser, the Burlington Northern Railroad, and the Washington Department of Natural Resources. There were also a number of private mining claims dating back to the turn of the century. The volcano was in the jurisdiction of the U.S. Forest Service, but there was much land owned privately, as well as by Cowlitz and Skamania Counties and by the state of Washington.

It was in this wondrous natural playground that the Mount St. Helens eruption began so inauspiciously on March 20, 1980, and erupted so violently on May 18.

Tracking the Eruption

The eruption began with a magnitude 4.1 earthquake at 3:37 P.M. on March 20, 1980, with no visible external ash plumes or steam vents. After March 20, small earthquakes continued, their cause being the rise of magma through the conduit beneath and into the volcano. At an early stage, the U.S. Geological Survey (USGS) began assembling a team of geologists and volcanologists and moving them to Vancouver, Washington. Many of the volcanologists with the USGS had learned their craft at the Hawaiian Volcanological Observatory, and through the years, became part of a large group having direct experience with an active volcano, although that volcano was much less explosive than the Cascade composite volcanoes. Other geologists, those with expertise in floods and mudflows, were also called. Some of the personnel were moved directly from the Hawaiian Volcanological Observatory. The geoscientists immediately began to establish a network of seismographs around the volcano to monitor the rise of the magma. A system to detect small movements of the volcano was also set up. Tiltmeters can detect ground tilts and deformation of very small magnitudes, on the order of hundredths of a millimeter. If the instruments are used on all sides of the volcano, they can record balloonlike swelling, which is caused by magma flowing into the volcano from below. Methods that detect volcano swelling have been used successfully for years in Hawaii to forecast imminent eruptions.

The first visible eruption of Mount St. Helens occurred in the early afternoon of March 27; but on March 24, before there was any explosive volcanic activity, geologists from the University of Washington and personnel from the U.S. Forest Service began advising the public to stay away from Mount St. Helens and from Spirit Lake. On March 25, the forest areas above Timberline, a camping area on the southern flank of the volcano, and the Forest Service's information center at Spirit Lake were closed. The Skamania County sheriff's office closed State Highway 504 at a point about 8 kilometers northwest of the volcano's peak on the same day. By this time, news reporters and photographers had flocked to the mountain, and airplane traffic was so heavy that the Federal Aviation Administration (FAA) imposed special flight restrictions.

The first meeting of federal, state, and county emergency-services officials was held on March 26. Les Nelson, the sheriff of Cowlitz County, was skeptical when he received a call to attend the meeting. To illustrate his point that officials and organizations knew little about one another prior to the meeting and that they soon learned to respect each other, Les Nelson humorously told of his frame of mind prior to that first meeting. He stated, in a talk at a workshop on volcanic hazards in Sacramento, California, in December 1981:

> The Forest Service was going to be there. I've heard about them; They're the guys that wear Smokey Bear hats and tell us to prevent forest fires. DES (Department of Emergency Services) was going to be present. They're the folks up in Olympia that come down once a year to tell hunters not to get lost. The USGS was going to be there. Well, I remember seeing a picture of those guys in NATIONAL GEOGRAPHIC once. They're the guys who run around with little hammers pecking at rocks and making notes in books. The Department of Transportation was going to be there. They're the guys that drive around in yellow trucks all day picking up paper and looking through a transit at each other. The County Commissioners—I was familiar with them; they're the people who give you a little bit of money and then give you heck about the way you spend it. The National Guard—they're the guys that have the bunch of kids that dress up like soldiers on weekends; they carry bayonets and look like a walking rummage sale. I don't know what they do—just have fun, I guess. State Patrol—we are familiar with them, naturally; they're the guys that drink coffee while we cover their accidents. (Nelson, "Local government's experience," p. 81)

In the March 26 meeting with many emergency-oriented organizations, participants of the meeting heard the geologists talk about earthquake activity and the nature of the potential hazards. As a result of the meeting, an emergency coordination center was set up by the Forest Service at its Vancouver headquarters. This decision greatly assisted newly launched scientific efforts and was instrumental in providing anxious reporters, officials, and the public with hazard warnings and information.

In spite of the earthquakes and news reports, some local people remained suspicious of the officials' motives, and one cynic said that the alarm of a possible eruption had been invented by the Forest Service to delay the development of a recreation area at Spirit Lake. One official likened the planning situation to building a boat while paddling it upstream. But since the volcano was mostly on lands administered by the U.S. Forest Service, it was natural for them to take an early and permanent role in coordinating efforts. They had ample experience in dealing with forest fires in the region, so they knew the terrain and had an extensive radio network around the volcano.

There had been a week of bureaucratic activity concerning scientific studies and emergency planning after March 20, but the steam explosions on March 27 galvanized officials into even greater action. Volcanologists from the USGS began intensive measurements of volcanic gases, searching for evidence of sulfur dioxide, an indicator that fresh magma is rising. They intensified their studies of volcanic particles—searching for fresh bits of magma—and established more stations to measure ground deformation. The USGS issued a formal "Hazards Watch" to more than three hundred state and federal officials, to representatives of other agencies, and to a local congressional delegation.

Within hours after the first explosive activity of March 27, hundreds of people were evacuated from logging camps, mountain homes, and public facilities. Emergency-services officials advised residents within a 24-kilometer radius of the mountain to leave, a distance that would prophetically match that of the farthest-traveled blast cloud during the eruption of May 18. Forest Service employees and their families left the ranger station at the head of Swift Reservoir. About three hundred loggers were moved out of three Weyerhaeuser Company logging camps near the volcano, as were twenty people from the state fish hatchery on the North Fork of the Toutle River, about 48 kilometers downstream from the volcano. Skamania County Sheriff's deputies moved forty-five people, mostly newsmen and geologists, from the Spirit Lake area. Cowlitz County law enforcement officers evacuated people farther downstream, along the Toutle River. Deputies from both counties set up roadblocks on several main routes to keep out the curiosity seekers.

In spite of their efforts to create a reasonable buffer zone based on then current scientific expertise, officials were hounded by countless people who felt that they had a right to pass into the red zone. These included loggers whose livelihood was jeopardized, owners of cabins and homes and other land parcels, and the press. The Cowlitz County sheriff, Les Nelson, reported a wide range of disgruntled comments, such as: "This is a free country." "This is a public road." "I'm a property owner; I have a right to go up to my property and get my earth stove out of there." "I pay taxes on that set up." Store owners were unhappy. Hunters were unhappy. Fishermen and tourists were turned away. As a result, latent suspicions of government bureaucracy became highly evident.

It proved impossible to keep everyone out of the danger zone because of the interconnecting network of logging roads that laced the area, especially in the region north of the volcano where the Weyerhaeuser Company was actively logging timber from its own land. The sheriff of Skamania County, Bill Closner, reported that it didn't matter what officials did. People went around, through, and over barricades. Law enforcement officials received continual verbal and written abuse, some of it from cabin owners at Spirit Lake. Harry Truman absolutely refused to leave his home and rental cabins at the south end of Spirit Lake. Just before the eruption on May 18, when volcanic activity had temporarily subsided, people with property or relatives in the area that was later devastated became increasingly vocal in their demands to return. Despite the heavy pressure, the efforts of law enforcement officials and the Forest Service would pay off in the end. Their vigilance and the fact that May 18 was a Sunday and the loggers were not at work would account for the fact that not more than fifty-seven lives would be lost. Had access not been limited, several thousand fatalities might have occurred.

The placement of roadblocks illustrates the struggle between the volcanolo-gists, who understood the dangers of the volcano, and people who were critical of authority and personally involved with immediate economic and social con-cerns. Yet many disaster-related decisions, such as the placement of road-blocks, were decided by nongeological factors. In the first few weeks after the beginning of volcanic activity, the first roadblock was established at Camp Baker for volcanological safety reasons (see fig. 1.4). For economic reasons, however, it was moved to the county line (position 2) so that Cowlitz and Skamania Counties could jointly man the roadblock and share expenses. Then for disaster-related reasons it was shifted back to position 3—the volca-nologists said position 2 was too close to the danger zone. However for eco-nomic reasons, the loggers at Camp Baker, which was operated by the Weyer-haeuser Company, put pressure on emergency-response officials to move it back to Camp Baker so that they could return to work. But at this location the crowds interfered with the loggers' access, so the roadblock was shifted farther back to location 5, which was at the junction of Highways 505 and 504. The choice of this location was not influenced by the possibility of volcanological hazards.

Similar blockade shifting occurred south of the volcano. The first roadblock was placed 4.8 kilometers east of the small town of Cougar but was then pushed up to the county line so that Cowlitz and Skamania Counties could share the expense. Then volcanologists advised that it be located farther away, so it was moved back to a rural store (Jack's), a move strongly endorsed by the owner for reasons of increased revenue. The people of Cougar then com-plained that they had lost revenue, so the blockade was returned to the county line, and then back to its original position 4.8 kilometers east of Cougar. The Washington Department of Emergency Services distributed information sheets to the residents of Cougar suggesting that they pack overnight bags and be ready to evacuate quickly. This action only angered the people because of the presumption that officials might try to remove them from public and pri-vate property.

The volcanologists were becoming increasingly concerned, for on March 28, at least a dozen eruptive episodes occurred and the earthquake activity continued. But the most disturbing event was silent and invisible—sulfur diox-ide was found in the gases emanating from the crater, indicating that the erup-tion plume was more than just superheated water—there were products of live magma present. With this discovery, the warnings grew more ominous, al-though suspicious sightseers regarded them as scientific overreaction. Every possible evacuation route became jammed with arriving vehicles, and many tried to evade roadblocks to get closer views. Officials of the Washington Department of Emergency Services beseeched the public to stay away from the potentially dangerous areas, but the nonbelievers and the reckless ignored their pleas. On the afternoon of March 30, seventy aircraft were reportedly flying around the volcano at the same time, engendering strict air-traffic controls to prevent collisions. On March 31, some people in helicopters landed on the crater rim and climbed inside. One group with camouflage clothing climbed to the top and filmed scenes intended for a documentary movie and for beer commercials.

Meanwhile, the initial steam explosions of March 27 had blown out a small crater, scattering dirty gray volcanic ash and dust across the once clean white snow and ice at the summit, and on March 29, the explosion crater had grown larger, as earthquakes continued. On April 1, the extensive seismic network detected harmonic tremors below Mount St. Helens. Such tremors are the vibrations associated with movement of magma, analogous to those generated by fluid moving through a pipe. These tremors, in conjunction with escaping sulfur dioxide, convinced volcanologists of the high probability of an imminent eruption.

Although explosions were occurring, volcanologists and officials still had trouble convincing a skeptical public that the volcano was dangerous. On the same day that the harmonic tremors were detected, about 300 loggers fighting to return to work won a release to return to their job sites on the lower flanks of Mount St. Helens. Despite warnings, illegal entries were rampant, but there were not enough personnel to prevent all of them. County budgets were too small to hire more people. In desperation, officials of Cowlitz and Skamania Counties asked the Washington National Guard to help maintain the roadblocks around the volcano. About 180 officers would be needed to adequately maintain an area that included the network of logging roads and twenty-nine barriers. Dixie Lee Ray, then governor of Washington, declared a state of emergency and set up the Mount St. Helens Watch Group on April 3. On April 4, she authorized the National Guard to share duties around the mountain, and on April 5, 60 guardsmen took up their posts at four roadblocks. Only property owners and scientists were allowed inside the restricted area, but the loggers continued exerting great pressure on county officials for access to the camp.

John Sorenson of Oak Ridge National Laboratory, who studied emergency planning for such rare events as volcanic eruptions, read the lead articles of several local newspapers each day from March 21 to April 11 and recorded the conflicting information being printed. On some days it was reported that an eruption was likely, but on the next day it wasn't clear. One article even said that it was certain that the volcano would stop erupting in the near future. Some articles said that volcanic ash was no problem, and some said it could create breathing problems. The opinions bore no resemblance to the official information being distributed by the volcanologists who were on the official response team; the news was apparently the reporters' conceptions of what was happening. The overwhelming number of articles were about how many people were converging on the volcano and about the locations for the best views. Considerable attention was given to people, like Harry Truman, who were defying both the risk and the authorities. The ambivalence fostered by the press created skepticism in the minds of the public and tended to destroy the credibility of the scientists, thereby hindering efforts to save lives.

By April 12, Skamania County officials expressed concern about how long they could control access to the mountain. Some questioned whether it was appropriate for them to keep people out of the area against their will. A mutiny of the public seemed imminent. Public officials and volcanologists were accused of overstating the dangers, and many citizens believed that the eruption was nearly over and that the volcano was no longer dangerous.

At the same time, there was no stopping the progress of the eruption. Steam explosions continued, and the summit crater was progressively enlarged. Harmonic tremors increased, and on Sunday, April 13, there were eighteen eruptions of steam and ash. Clear weather brought record-breaking numbers of sightseers to the area from all over the United States and from other countries. On the weekend, two television camera crews were airlifted onto the mountain, and there were unauthorized climbers. All were in defiance of closure orders. On April 14, eruptive activity had decreased. Even so, the Washington Department of Game gave notice that three popular fishing lakes near Mount St. Helens were to remain closed. There was a predictable outcry from fishermen eager for early spring fishing.

It was announced on April 17 that the greatest potential hazard was the increasing instability of the northern flank of the volcano, and on April 19, officials reported that part of the upper northern side of the mountain had begun to bulge outward. Most volcanologists took this to mean that magma was pushing into the volcano and causing the protrusion. The potential dangers associated with the bulge were debated by the geologists. Arguments centered upon how far the destruction might extend if an eruption occurred. Some scientists expected large landslides. Some argued that large pyroclastic flows would extend across the mountains and produce a horizontal blast as had happened at Bezymianny Volcano, in Russia, in 1956. As the swelling on Mount St. Helens grew, the potential for extended catastrophe also increased, but the catastrophe that materialized on May 18 went far beyond what had been forecast by the official announcements at that time. In other words, no single geologist knew what was going to happen—some believed that the extent of the eruption would be limited to the North Fork Toutle River valley, whereas others believed that it would extend much farther in a gigantic explosion. There was no consensus; instead there was constant controversy. All of the geologists had an opinion, and they all knew that the potential for disaster was large. As the bulge swelled, the volcanologists became increasingly alarmed, for the decline in eruptive activity continued at a time when the probability of an eruption was increasing. This lull in explosive activity, which occurred between April 14 and April 23, seduced many local people and reporters into believing that the eruption was waning, while at the same time the volcanologists were forecasting possible disaster. On April 23, some volcanologists warned that if the bulge continued to grow, the unstable slope could cause large avalanches that would engulf the shores of Spirit Lake.*

The bulge continued to grow, so Governor Ray and the Forest Service supervisor closed additional areas near the volcano on April 30. The red zone was established around the peak at distances ranging from 4.8 to 16 kilometers and included Spirit Lake. Only scientists, law enforcement officers and other officials, and search-and-rescue personnel were permitted in the zone. A blue zone signifying a smaller-risk hazard zone surrounding the red zone was also established. In this new zone, logging operations were allowed to continue, and property owners holding special permits could enter during daylight hours. At the same time, the governor declared the entire state an emergency area because volcanic ash could cause damage anywhere in the Pacific Northwest.

On May 1, a new observation point was established at 9.2 kilometers northwest of the volcano on a ridge that directly overlooked the mountain. The lookout was called Coldwater II and consisted of monitoring instruments and a house trailer that was in radio contact with headquarters in Vancouver. The observation post was necessary to monitor the growth of the bulge as well as give visual early warning of mudflows and avalanches. Until May 17 it was manned solely by Harry Glicken, a graduate student at the University of California, Santa Barbara, and summer field assistant for David Johnston, a young researcher for the USGS. The post was placed on the ridge to allow more frequent monitoring of the bulge. The rate and amount of northward movement of the bulge had the response team concerned that measurements were not being taken often enough to catch the signal of an impending landslide. They gambled that the information gained from the manned observation post would help protect the public. Yet, it was thought to be in a dangerous location. Prior to the day of the eruption, plans had been made to move a half-track vehicle to the post on May 18 for better protection of personnel. The magnitude of the blast, however, would exceed the official forecasts, which were based upon a previous eruption that occurred about 1,200 years ago and had produced a blast that traveled a distance equivalent to Spirit Lake. Hence, the half-track probably would have had little or no protective value.

---

* The uncertainty in forecasting an imminent catastrophic eruption at Mount St. Helens in 1980 deserves comment and points out the immense strides that have been taken since then. In 1980, forecasting tools that worked in Hawaii were being applied to an explosive composite volcano with which there had been little experience in the United States and for which there had been few examples described in the volcanic literature. It took nearly ten years of additional study of explosive volcanoes plus development of fast, high-capacity, portable computers and microelectronics to arrive at the forecasting capability that achieved notable success in saving thousands of lives during the 1991 eruption of Mount Pinatubo in the Philippines (see chap. 15). That knowledge and technology did not exist in 1980.

After witnessing the steady swelling of the bulge on the north side of the volcano, why didn't the geologists forecast that a blast was coming? The answer is that some did and some didn't, but no one definitely knew. In 1956 the Russian volcano Bezymianny exhibited behavior similar to that of Mount St. Helens prior to eruption. Many of the geologists were aware of the Bezymianny event, and the article detailing the eruption was available to all of them. Norman Banks, one of the USGS volcanologists on duty before the eruption, commented that "Bezymianny did dominate more than one evening meeting, when strongly posed arguments fell on both sides of whether a Sugar Bowl [a prehistoric small lateral blast followed by dome building] or a Bezymianny event was in store. [The Bezymianny event was a large lateral blast.] I believe that David Johnston was one of the pro-Bezymianny proponents" (Banks, e-mail to RVF, June 6 and 24, 1996; interpolations in original).

Jack Hyde, a geology professor at the Tacoma Community College in 1980, who was not part of the crisis team, was interviewed by Jim Erickson of the *Tacoma News Tribune*, which published an article on May 6, 1980. The headline of the article was "Geologist: Bulge may pop like lava balloon." The article, verbatim, follows:

> The growing bulge on the upper north slope of Mount St. Helens is "the most significant" development since the volcano began erupting on March 27, says Tacoma geologist Jack Hyde.
>
> Based on his extensive research of St. Helens and his knowledge of other volcanoes, Hyde said he believes the bulging is caused by lava moving beneath the surface of the 9,677-foot mountain.
>
> Hyde said he has no idea how long the peak could continue to bulge at its present rate of about 5 feet a day before something happened.
>
> "I have a gut feeling, though, that as the bulge continues to grow, something dramatic is going to happen soon," Hyde said.
>
> He speculated that as the north slope becomes more unstable and steeper due to the bulging, it will cause massive landslides.
>
> What could follow is a "spectacular" explosion of lava into the air as lava vents on the north slope are opened up. This explosion, he added, could come without warning.
>
> "It could come suddenly without any leakage of magma gases," Hyde said. Such gases, with a high concentration of sulfur dioxide, have preceded lava eruptions on other volcanoes.
>
> "But St. Helens doesn't have a free vent like volcanoes in Hawaii," Hyde said. "There are a lot of volcanoes on record that have cleared lava vents in huge explosions."
>
> "Mount Pelée on Martinique in 1902 blew the lava plug from a crater vent, pushing lava 1000 feet into the air," Hyde noted.
>
> "And the blast from the explosion of a Soviet volcano in 1956 blew down trees 15 miles away," he added.

He said Mount St. Helens could do likewise.

Told that scientists are watching the mountain from ridges near the peak, Hyde replied: "I hope they're not in a direct line. That's like looking down the barrel of a loaded gun."

In hindsight, we can label Jack Hyde, as well as many members of the Mount St. Helens crisis team, as being prophetic because now we can clearly filter out all of the other considerations and uncertainties of the time. Decisions and forecasts are taken from data, information, experiences, and interpretations of many participants who have worked on many volcanoes, and Bezymianny did not stand out, among the many others that were considered, as the most obvious volcanic event that would occur at Mount St. Helens. To correctly assess what happened, one must relive the tensions and uncertainties of the time to understand why certain decisions were made and not point a finger just at failures that can only now be recognized because of what eventually happened. The scientists who were in charge did an admirable job. Daily they crafted or revised official releases from the results of each evening's discussion sessions, in which five to twenty volcanologists reported their findings and expressed strong opinions not only about their own data and ideas, but also about everyone else's. Agreement was never unanimous, which is to be expected. To illustrate the impact of hindsight on how we view history, imagine what we would say today if the eruption had not taken place. Those who had argued against an eruption would now be acclaimed as being trustworthy and of good judgment.

As fate would have it, Harry Glicken had made an appointment in the previous February with one of us (RVF) to be in California on May 18 to discuss his future graduate studies at the University of California in Santa Barbara. He was therefore given permission to leave for his appointment on the afternoon of May 17, and his duties were voluntarily assumed by David Johnston, his immediate supervisor. The following day, at 8:32 A.M. on May 18, the mountain collapsed and a blast far exceeding anything that anyone expected blew out the north side of the volcano in the approximate area of the bulge. As the dark gray turbulent mass hurtled northward, the last words of David Johnston were "Vancouver. Vancouver. This is it!"

THROUGHOUT the latter half of the twentieth century, substantial funding for the sciences in many countries, including the United States, fostered extensive growth of basic scientific research in physics, chemistry, biology, and the earth sciences. This work led to unexpected discoveries that currently benefit the health and well-being of humankind. It is the unexpected discoveries that distinguish basic research from applied research, which uses those discoveries for new applications. In the United States, generous federal funding for the earth sciences provided a great opportunity for research using seagoing vessels and new remote sensing equipment to study the seafloor with no other purpose than to learn more about how the earth was made. These investigations have led to a better understanding of how the earth works and why volcanic eruptions occur, which we explore in the next chapter. Much of the enhanced ability to save lives from volcanic eruptions comes from what has been learned in the past forty years.

References

Blong, R. J. *Volcanic Hazards: A Sourcebook on the Effects of Eruptions*. Sydney, Australia: Academic Press, 1984.

The Daily News (Longview, Washington) and *The Journal-American* (Bellevue, Washington). *Volcano: The Eruption of Mount St. Helens*. Longview, Wash.: Longview Publishing Co.; Seattle, Wash.: Madrona Publishers, 1980.

Erickson, J. "Geologist: Bulge may pop like lava balloon." *Tacoma News Tribune*, May 6, 1980.

Foxworthy, B. L., and M. Hill. *Volcanic Eruptions of 1980 at Mount St. Helens: The First One Hundred Days*. U.S. Geological Survey Professional Paper 1249. Washington, D.C., 1982.

Lipman, P. W., and D. R. Mullineaux, eds. *The 1980 Eruptions of Mount St. Helens*. U.S. Geological Survey Professional Paper 1250. Washington, D.C., 1981.

"Mount St. Helens Diary: A Sunday Holocaust." *Grants Pass Daily Courier*. 1980.

Nelson, L. "Local government's experience at Mount St. Helens." In *Status of Volcanic Prediction and Emergency Response Capabilities in Volcanic Hazard Zones of California*. California Department of Conservation, Division of Mines and Geology Special Publication 63, 3-81–3-87. Sacramento, Calif., 1981.

Pringle, Patrick T. *Roadside Geology of Mount St. Helens National Volcanic Monument and Vicinity*. Washington Department of Natural Resources Circular 88. Olympia, Wash., 1993.

Saarinen, T. F., and J. L. Sell. *Warning and Response to the Mount St. Helens Eruption*. SUNY Series in Environmental Public Policy. Albany: State University of New York Press, 1985.

CHAPTER 2

Why Do Volcanoes Erupt?

Frontispiece. Volcanic eruptions can affect history. This turn-of-the-century stamp depicts Momotombo Volcano, Nicaragua, in eruption. Plans were being made in 1902 for a canal across Nicaragua, linking the Pacific with the Gulf of Mexico, but were dashed by a combination of news about 29,000 deaths from the eruption of Mount Pelée, and the issuance of a stamp bearing an image of the smoldering Momotombo. With the Mount Pelée catastrophe fresh in mind, opponents of the Nicaraguan route prominently displayed the stamp during a U.S. Senate debate. The proponents of a Panama route gleefully exaggerated the potential effects of Nicaraguan volcanoes on a shipping route across Nicaragua. As a result, the route across what is now Panama was chosen by American investors.

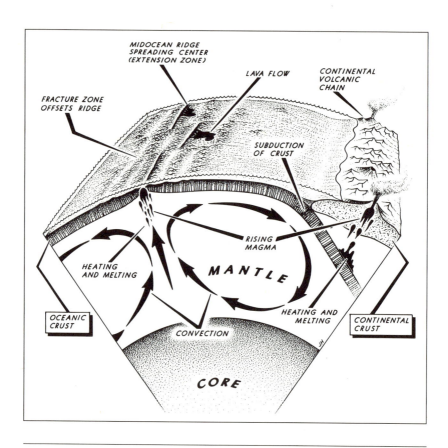

Fig. 2-1. Mount St. Helens and the other Cascade volcanoes of the Pacific Northwest overlie a zone where slowly moving oceanic crust is pushed beneath the thicker crust of the North American continent. The engines that drive this movement are rising heat currents (convection currents) in the mantle below the crust. Along this subduction zone, the thinner crust, composed of ocean-floor lavas and sediment, is pushed down to a depth where higher temperatures and pressures partially melt the rocks. The molten material (magma) rises buoyantly toward the earth's surface. If enough of the magma collects into larger masses and the zone is heated, a small portion reaches the surface to erupt. At Mount St. Helens, the rise of magma could be followed by watching depths of small earthquakes with a network of seismographs and by the bulging volcano surface.

We know the earth as the "blue planet." Photographs taken by astronauts from space show that the earth has a distinct blue color because so much of its surface is occupied by water; it is the only blue, solid planet in our solar system. But what do we find by looking into the earth? Seismic instruments recording earthquake data and deductive reasoning from what we know about other planets and matter within the solar system reveal that the earth has a core, an intermediate zone known as the mantle (*aesthenosphere*), and a crust (*lithosphere*). The core is believed to be a mixture of nickel and iron, the mantle is semi-solid rocky material, and the crust is brittle rock.

There is a tremendous difference between the temperature of the core and that of the crust. We are fairly certain that this difference sets up convection currents as the hot mantle rises, cools down when it nears the crust, and then sinks, only to heat up and rise again. It is believed that this dynamic response to heat and density differences has caused the brittle crust to break up into a mosaic of crustal plates (*tectonic plates*) that ride on the upper parts of convection currents, pulling the crust apart on one end and pushing it together on the other end (fig. 2-1).

The melted rock that forms magma does not come from the earth's core or from deep within the mantle. Nor is there evidence of permanent pools of melted rock. Magma is formed in the upper part of the mantle or within the crust mainly at tectonic plate boundaries (fig. 2-2). Where plates move together, one end is pushed beneath the other to form *subduction zones*. As the Pacific plate pushes against the American plate, it descends (is subducted) and heats up. One theory is that part of the subducted plate gets so hot that it melts and the melt then rises into the crust to erupt as volcanoes. Since the subducted crust is shaped like an elongate slab, volcanoes occur in lines above its edge. The volcanoes of the Cascade Mountains, including Mount St. Helens, lie above a subduction zone.

Sudden slippage between tectonic plates can cause earthquakes. Simultaneously, it can cause frictional heat, adding to the high temperatures within the earth. These processes combine to melt solid material to form local pockets of magma, which then rise buoyantly, like slow-moving balloons, into the crust to form the pools of magma that feed volcanoes. The pools of magma are sometimes called *magma chambers*, although there is no empty cavity as the word *chamber* implies. Along plate margins, earthquakes and volcanoes have a common cause and therefore a similar distribution around the earth. While

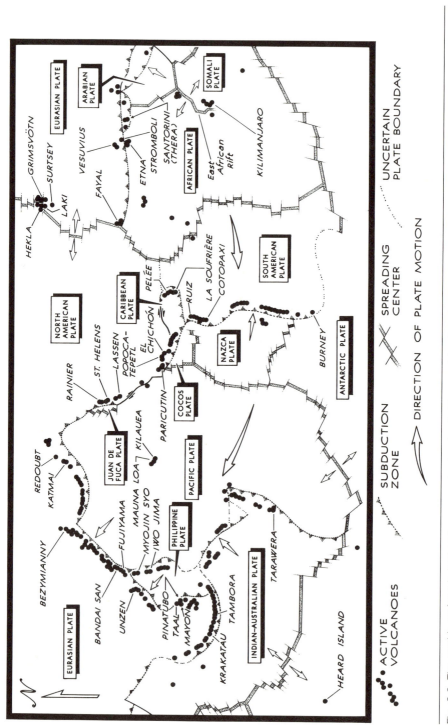

Fig. 2-2. The earth's surface is broken by zones of extension (midocean ridges) and collision (subduction zones) between crustal plates. The relatively stable portions of the crust between these zones are called *crustal plates*: volcanism and earthquakes occur at their margins: New crust is made along midocean ridges by rising basaltic magmas erupting along and filling the widening cracks, and crust is thickened over subduction zones.

we know some generalities about magma formation and its rise to the surface to erupt, we still do not know the precise reason why a volcano erupts at a specific place and a specific time.

The designation Ring of Fire refers to the discontinuous lines or chains of volcanoes (*volcanic arcs*) that occur around the Pacific Basin. The volcanoes that form the Ring of Fire erupt above subduction zones of the Pacific plate, which is moving toward the continental and oceanic plates that surround it. Starting with the Aleutian Island volcanic arc in Alaska, chains occur through western Canada, along the Pacific coast of North and Central America, through the Andes, and even farther south into the Antarctic continent. More complicated volcanic chains and arcs also run north along the western Pacific Ocean margin, from New Zealand, through Indonesia, the Philippines, Taiwan, Japan, Kamchatka, and back to the Aleutian Islands in Alaska, completing the ring.

Where plates move apart, hot mantle material rises into long fractures and causes the crust to grow. These areas, *extensional boundaries*, occur between plates on the floors of major oceans, for example, along the mid-Atlantic ridge, and separate large segments of the earth's crust. As the crust is fractured and moves apart, molten rock or heat finds easy access to the surface. Impressive as the world's visible chains of volcanoes are, it may surprise some readers to learn that most of the earth's volcanic activity is hidden beneath the sea along these seams between crustal segments.

Not all volcanoes occur at subduction zones or at extensional plate boundaries. Some areas of volcanism, such as the Hawaiian Islands and the Yellowstone Plateau, are within the main body of a plate, well away from its boundaries. For want of a better name, they are said to be built above *hot spots*, which are believed to be thermal plumes rising through the mantle. Some thermal plumes have existed for millions of years in about the same place, and the crust has moved over them.

Buried under enormous masses of rock, all materials beneath the earth's surface are under great pressure, and it is this pressure that keeps gases in solution within magmas. Magmas that are low in gas content or have properties that prevent gas from escaping will flow out onto the earth's surface quietly as lava flows. Magmas that rise quickly to the surface with abundant gas in solution experience rapid pressure drops leading to explosive expansion. The magma disintegrates to form partly-molten to solid *pyroclasts* (volcanic ash); these particles can be carried in eruption clouds to as high as 50 kilometers into the sky.

The eruptions at Mount St. Helens and Mount Pelée were the final episodes in a process that started millions of years earlier as newly formed oceanic crust slowly collided with another crustal plate and was thrust downward to a depth at which some of the rock melted. Bits of melt accumulated into bodies large enough that slow buoyant rise was possible for the aggregate. A small amount of the molten mass finally penetrated the earth's crust, reaching low-pressure regions at the earth's surface, where it erupted. At both Mount St. Helens and Mount Pelée, magma rising from great depths and carrying gases in solution, as well as crystals floating in the molten mass, may have followed fractures upward until the pressure within the magma body exceeded the pressure caused by the weight of overlying rocks. At both volcanoes, the magma fragmented explosively and produced pyroclastic flows (see chap. 5) and ashfalls. Each of these volcanic eruptions, as well as many others, has provided data and experience used to design more refined hazard zonation and better mitigation strategies at eruptions elsewhere. Thousands of people at Pinatubo, Philippines, for example, were saved in 1993 (see chap. 15) on the basis of knowledge gained over the last century, especially during the last forty years.

From the Tiny Molecule, Giants Grow

Rocks are clusters of minerals and usually contain several varieties. Each of the minerals is a chemical compound, usually crystalline, composed of one or more elements, and formed in nature. Minerals have distinctive physical and chemical properties, and those that form the earth's rocks can crystallize only from elements that are available in a magma melt.

Tables 2-1 and 2-2 show the elements and minerals in the mountains we climb and the ground upon which we walk and build our dwelling and roads. These elements form the rocks we collect and the substance of many construction materials. Most of the atoms in the rock-forming minerals of the earth's crust are bonded to the silicon-oxygen tetrahedron molecule. Such minerals are called silicates. The silicon tetrahedron is necessary to life because its structure determines the framework of most of the solid parts of earth, including the sizes and shapes of the world's volcanoes, and strongly influences the rates of erosion and chemical breakdown of rocks that release the elements necessary for the growth of life-giving plants.

The basic building block of silicate minerals is the silicon-oxygen tetrahedron (fig. 2-3). Each tetrahedron is composed of four oxygen atoms and one silicon atom and has four negative charges, which attract positively charged atoms to the corners. The oxygen atoms are located at the apexes of the tetrahedron, and the much smaller silicon atom resides within the center. These tetrahedral units combine with the other elements shown in table 2-1 to form silicate minerals.

Crystallization of silicate minerals involves the bonding of silicate tetrahedra to one another in increasingly complex patterns, which include chains, rings, and three-dimensional networks. Silicate minerals do not normally form in isolation, but rather crystallize from silicate soups—magma—containing all of the elements from whatever part of the crust or upper mantle that melted to make the magma (table 2-3).

**Table 2-1. Dominant Elements
of the Earth's Crust (%)**

Oxygen (O)	46.40
Silicon (Si)	28.15
Aluminum (Al)	8.23
Iron (Fe)	5.63
Calcium (Ca)	4.15
Sodium (Na)	2.36
Magnesium (Mg)	2.33
Potassium (K)	2.09
All others	0.66
	100.00

**Table 2-2. The Most Common Minerals in
the Earth's Crust**

Minerals (%)	*Elements in the Minerals*
Feldspar (51)	
Orthoclase	K, Al, Si, O
Plagioclase	Ca, Na, Al, Si, O
Quartz (12)	Si, O
Pyroxene (11)	Fe, Mg, Al, Si, O
Mica (5)	
Biotite	K, Fe, Mg, Al, OH, Si, O
Muscovite	K, Al, OH, Si, O
Amphibole (5)	Ca, Na, Mg, Fe, Al, OH, Si, O
Olivine (3)	Mg, Fe, Si, O
All others (13)	

Fig. 2-3. Most crustal minerals are silicates, with their basic building block being the silicon-oxygen tetrahedron, a three-sided pyramid, with its base being the fourth side. The concentration of silica building blocks determines the property of the fluid magma. Magma with 55 percent or less silica is less viscous than magma with more than 70 percent silica. The viscosity of magma, and ultimately the submicroscopic silica building block, determines rock types and the shape and size of the mighty volcanic landforms.

Table 2-3. Common Volcanic Rocks and Volcano Forms

Volcanic Rock	Dominant Minerals	Fluidity and Silica Content	Dominant Volcano Form
Rhyolite	Quartz; Potassium Feldspar	Very viscous; Silica over 70%	Domes; Calderas
Andesite	Sodium Feldspar; Pyroxene	Moderately viscous; Silica 70%–55%)	Composite volcanoes
Basalt	Calcium Feldspar; Pyroxene	Very fluid; Silica under 55%	Cinder cones; Shield volcanoes

To enhance communication, the names given to rocks like those in table 2-3, are a convenient way for geologists to compress a large amount of background information into a single word. For example, by using the single word *rhyolite*, the geologist is saying, "This rock is made mostly of microscopic quartz and feldspar minerals, with a chemical composition having a high percentage of silica and potassium, which in the liquid form has a very high viscosity. It is a volcanic rock."

Magma containing less than 55 percent silica (*basalt*) flows easily whereas magma with more than 70 percent silica (*rhyolite*) flows with difficulty and is more viscous than cold tar. With the highly fluid, low-viscosity basaltic lava, the rapid escape of gas produces fountains of incandescent lava blobs and drops. The latter forms are often fluid enough when they hit the ground to coalesce and form lava flows. But with the high-viscosity rhyolite lava, gas cannot readily escape unless the pressure is high enough. If it is, the gas may escape suddenly and violently, and the mass flies apart into an explosion of glassy bits of quickly cooled magma (glass shards) and crystals. Glass shards are the main component of many deposits of volcanic ash (fig. 2-4). Sometimes the gases expand quickly to form a spongelike, glassy, light-weight mass known as *pumice* after it solidifies. Pumice forms in the throats of volcanoes and breaks into chunks when expelled into the atmosphere.

Initially, the composition of erupting magma is exactly the same as that of the rock melted to produce the magma. Generally, magma that originates from beneath or within an oceanic tectonic plate will be low-silica basalt that forms low-viscosity (highly fluid) lava flows. If the magma originates from beneath or within a continental tectonic plate, it will be of the high-viscosity, high-silica kind that forms rocks by explosive fragmentation. If the magma forms in areas where continental and oceanic plates overlap, its silica content may be of intermediate composition, and the magma will form andesitic lava flows (see table 2-3) as well as the pyroclastic material of explosive eruptions.

Fig. 2-4. The widespread ashfalls from explosive volcanic eruptions like the one at Mount St. Helens consist of particles that range in size from sand to fine flour. The major component shown here is natural glass, chilled during the fragmentation of a frothy melt as it came up the conduit into the crater. The glass shards are fragments from the walls of the bubbles. Also shown are frothy bits of pumice, formed when gas bubbles were stretched by the flowing magma, then partly fragmented in the eruption. In addition to glass shards and pumice fragments, volcanic ash also contains minerals formed in the magma chamber and bits of rock torn from crater and conduit walls.

The different behaviors of fluids with widely different viscosities can be visualized by imagining steam escaping from vigorously boiling water on a stove and then mentally comparing it with vigorously boiling, higher-viscosity oatmeal. The oatmeal spatters explosively onto the stove, analogously to pyroclastic debris exploding from viscous magma around a volcanic vent.

The types of eruptions, the shapes, and the sizes of volcanoes ultimately depend upon the attributes of the silica tetrahedron because the amount of silica determines the viscosity of volcanic fluids, and the viscosity determines whether lava flows or pyroclastics are dominant during eruptions. These, in turn, determine the shapes and sizes of volcanoes (see chap. 3).

Volcano Country

Some people are concerned about living near volcanoes. In the United States, there are potentially active volcanoes in Alaska, Hawaii, the Cascade Mountains of Washington, Oregon, and California, the Rocky Mountain region around Yellowstone National Park, and near Flagstaff, Arizona. Such volcanic chains and clusters also occur in many countries around the Pacific Ocean and elsewhere in the world, potentially affecting the lives of millions of people.

The location of a volcano and the composition of the magma that feeds it determine the frequency and types of eruptions. Oceanic volcanoes fed by hot spots in the crust commonly have frequent nonexplosive eruptions of basaltic lava that often begin with spectacular lava fountains. Examples are the island of Hawaii, which is above a hot spot, and possibly Iceland, which is astride the mid-Atlantic Ridge. The Laki, Iceland, eruption of 1783 poured out 14 cubic kilometers of lava from a 25-kilometer-long fissure. Chemicals pumped into the air by that eruption created a European environmental disaster (see chap. 9).

High-standing volcanoes in chains such as the Cascade Mountains erupt less frequently but usually with more explosive violence than the basaltic Hawaiian volcanoes. Eruptions may occur only decades apart, but not always from the same volcano—individual volcanoes can be dormant for a few centuries at a time. Mount St. Helens has erupted approximately every 100 to 150 years since A.D. 1400. Mount Lassen, in northern California, last erupted in 1915. Crater Lake, in southern Oregon, was formed by a gigantic cataclysmic eruption about 7,000 years ago. Mount Baker in Washington erupted in the mid-1800s and showed increased fumarolic activity in 1975, causing apprehension that an eruption might occur. Warm spots and steam rise in the summit areas of Mount Rainier, Mount Shasta, and Mount Hood. Any one of those Cascade volcanoes could turn as deadly as Mount St. Helens did in 1980.

Extremely large eruptions occur far less often than small ones, because it takes a long time to build up the necessary enormous pressures within a magma body to produce a large eruption. At the extreme end of the scale, somewhere on earth every 100,000 years or so, an eruption occurs that spews forth an average of 10 cubic kilometers of material within a few days or at most, a few weeks. About 74,000 years ago, Toba, on the island of Sumatra, erupted 2,800 cubic kilometers of volcanic ash, over one thousand times more than the 1980 eruption of Mount St. Helens, which ejected 2.5 cubic kilometers of material! Toba ash covered huge areas of the Indian Ocean and Asia and may have altered the earth's climate, although as discussed in chapter 8, the abundance of ash may not perturb weather patterns as much as does the abundance of sulfur dioxide (fig. 2-5).

Over the five hundred years of record-keeping, it appears that eruption frequency has been increasing. This observation can be explained as an aberration

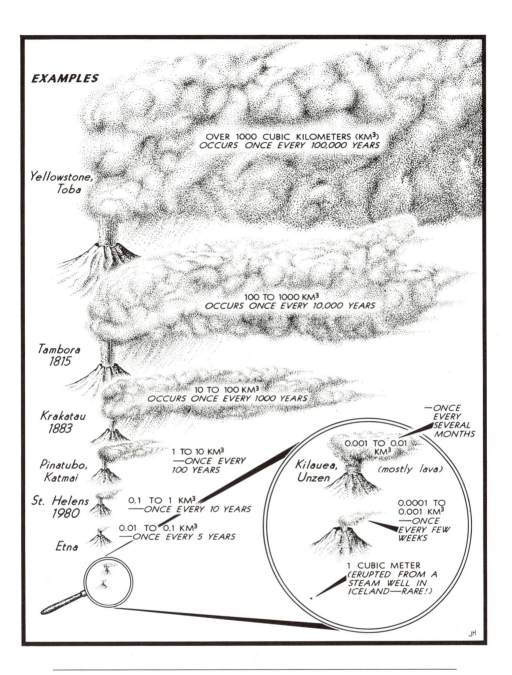

EXAMPLES

OVER 1000 CUBIC KILOMETERS (KM³)
OCCURS ONCE EVERY 100,000 YEARS

Yellowstone,
Toba

100 TO 1000 KM³
OCCURS ONCE EVERY 10,000 YEARS

Tambora
1815

10 TO 100 KM³
OCCURS ONCE EVERY 1000 YEARS

Krakatau
1883

—ONCE
EVERY
SEVERAL
MONTHS

0.001 TO 0.01
KM³

1 TO 10 KM³
—ONCE EVERY
100 YEARS

Kilauea,
Unzen   (mostly lava)

Pinatubo,
Katmai

St. Helens
1980

0.1 TO 1 KM³
—ONCE EVERY 10 YEARS

0.0001 TO
0.001 KM³
—ONCE
EVERY FEW
WEEKS

0.01 TO 0.1 KM³
—ONCE EVERY 5 YEARS

Etna

1 CUBIC METER
(ERUPTED FROM A
STEAM WELL IN
ICELAND—RARE!)

JH

Fig. 2-5. The energy or magnitude of an eruption is difficult to assess, but volcanologists often gauge it by the amount of erupted magma, which is determined by measuring the total volume of erupted lava and volcanic ash. Magnitudes shown in this diagram are from actual eruptions and are measured in increments of 10—from 0.000000001 cubic kilometer (1 cubic meter) to over 1,000 cubic kilometers—a difference of twelve orders of magnitude. The small magnitude eruptions occur much more often than the large ones—a fortunate inverse correlation for the earth and its living creatures. Eruptions with 0.001 to 0.01 cubic kilometers of material occur on the earth every few months. In stark contrast, eruptions of over 1,000 cubic kilometers of ash occur only once every 100,000 years. The frequencies of eruptions for the different volumes shown here are simplified estimates.

stemming from improved communication systems and increasing numbers of observers. It is also startling to discover that volcanic activity decreased during World Wars I and II, but this finding too, is an aberration with a simple explanation—not many people spent time reporting volcanic activity when all their efforts were dedicated to surviving a war. A complete tally of eruptions can also be impeded when activity is in uninhabited regions—these unseen eruptions are sometimes discovered later when a pilot flies over a volcano that is coated with fresh volcanic ash or that has a steaming crater. An unknown 1980 eruption of Gareloi Volcano, Alaska, was discovered only on archival meteorological satellite images when scientists were trying to explain anomalous sulfates in the upper atmosphere.

Awareness of volcanic activity has grown rapidly during the last two decades of the twentieth century as a result of several eruptions, including those of Mount St. Helens in 1980; Nevado del Ruiz, Colombia, in 1985; Mount Unzen, Japan, in 1991; and Mount Pinatubo, Philippines, in 1991. Increased thoroughness of reporting is largely due to modern technology—satellite observations, computer communications, and twenty-four-hour news networks. As the end of the twentieth century nears, the convergence of greatly increased reporting and computer analyses of eruption data is accelerating the knowledge of eruption frequencies and durations.

Of the thirteen hundred landlocked, potentially active volcanoes on earth, only a few—about a hundred—are monitored geophysically, and probably only half that number have been subjected to detailed study to evaluate their past eruption histories. Since each volcano is different, careful work is required to estimate how often and how long they erupt. By focusing on past eruption types and frequencies, perceptive advice can be given to planners in rapidly growing cities near volcanoes.

References

Blong, R. J. *Volcanic Hazards: A Sourcebook on the Effects of Eruptions*. Sydney, Australia: Academic Press, 1984.

Fisher, R. V., and H-U. Schmincke. *Pyroclastic Rocks*. Berlin: Springer-Verlag, 1984.

Francis, Peter. *Volcanoes: A Planetary Perspective*. New York: Oxford University Press, 1993.

Harris, S. L. *Fire Mountains of the West: The Cascade and Mono Lake Volcanoes*. Missoula, Mont.: Mountain Press Publishing Co. 1988.

Wohletz, K., and G. Heiken. *Volcanology and Geothermal Energy*. Berkeley: University of California Press, 1992.

Volcanoes and Eruptions

Frontispiece. Volcanologist rappels down crater wall of East Ukinrek Volcano, Alaska, to measure the thickness and volume of the layers, the age of the layers, and the kind of eruption that deposited each layer. The person on the rim holds the rope for the person in circle. Ukinrek erupted in 1977. Information from such descriptions helps to determine the frequency of eruptions and information about the destructiveness of past eruptions, and therefore a hint at what the future might bring. (Photo: Vicki Sue McConnell, Geophysical Institute, University of Alaska Fairbanks)

Fig. 3-1. The elegant symmetry of Mayon Volcano, Philippines, overlooking the Albay Gulf of Luzon Island and Legazpi City, belies the deadly eruptions that have occurred sporadically for hundreds of years. Volcanic cones like Mayon are constructed by tens to hundreds of eruptions of volcanic ash and lava and are called composite volcanoes or stratovolcanoes. Composite volcanoes can stand from less than 1,000 meters to perhaps as much as 5,000 meters above their bases. In the case of Mayon, its summit is 2,462 meters above the sea. (Photo: Arturo Alcarez, Philippine Commission on Volcanology)

An alien from outer space arriving on earth for the first time would most likely consider that all human beings, having the same basic structure, are alike. We all know that this is not so, and with familiarity, the alien would begin to see differences. Likewise the uninitiated will suppose that all volcanoes are alike, but in reality, they aren't.

A volcano is a mound, hill, or mountain constructed by the extrusion of lava or pyroclastic material, usually both, from beneath the ground. Even though individual volcanoes differ, they can be grouped into a few families according to shape, size, and type of volcanic material. Probing a bit deeper, volcanologists have learned that the composition of the original magma and the effects of its mixing with surface water are the principal reasons for the similarities and differences. Volcanoes within each family may resemble one another but they vary in physical details and eruptive behavior. A volcano's activity may change from time to time in an unpredictable way, and it is not easy to forecast eruptive behavior until it has been witnessed or determined by examining previous deposits. Even then, the forecast may not come true. In this chapter we will discuss the various members of the volcano family, such as composite volcanoes, domes, shield volcanoes, cinder cones, maars, and calderas. We will also explain the different types of eruptions, such as Strombolian, Plinian, and Hawaiian. Mount Pinatubo in the Philippines and Hawaiian volcanoes are discussed to illustrate some of the eruption processes.

Types of Volcanoes

Composite Volcanoes

Mount St. Helens and many other volcanoes of the world, including Mount Rainier and Japan's Mount Fuji, are graceful solitary cones known as *composite cones*, or *stratovolcanoes*. Their slopes are made of innumerable layers of rubble derived from the flow and break up of brittle lava and of dome rocks that are interspersed with some explosively produced pyroclastic layers and a few lava flows. Some of these volcanoes grow to greater than 3,000 meters above their bases (fig. 3-1). One of the highest in the world is Klyuchevskoy Volcano (in Kamchatka, Russia), which stands at 4,725 meters above sea level. Mount Fuji is 3,720 meters high. Depending upon the latitude, winter

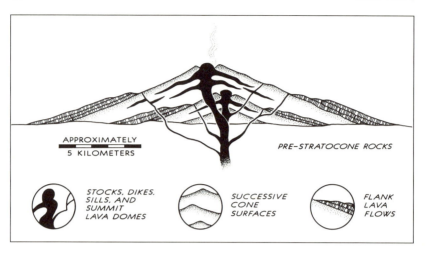

APPROXIMATELY
5 KILOMETERS

PRE-STRATOCONE ROCKS

STOCKS, DIKES,
SILLS, AND
SUMMIT
LAVA DOMES

SUCCESSIVE
CONE
SURFACES

FLANK
LAVA
FLOWS

Fig. 3-2. Diagram of a composite volcano. Repeated extrusions gradually build up a complex, layered series of cones around a vent or a series of closely spaced vents. Occasional intrusions of magma and lava flows contruct a framework that gives support to the accumulation of rubble.

snows whiten the slopes of these volcanoes from the top to the base, and the heat of summer melts all but the highest, most protected, or deepest snow fields. On the upper slopes of some volcanoes, such as Mount Rainier, perennial snows spawn glaciers that descend for many miles down the mountain as rivers of ice.

Most composite cones stand isolated several tens of kilometers apart in elongate chains such as those along the Pacific Ocean's Ring of Fire (see chap. 2). Composite volcanoes are constructed from multiple eruptions, sometimes occurring over hundreds of thousands of years, sometimes over a few hundred. Andesite magma, the most common but not the only magma type forming composite cones, is more explosive than basaltic magma. The andesitic eruptions provide abundant fragmental debris to construct the stratocones. However, not all of the magma rises to the summit of a volcano. Some penetrates through the layers of a volcano as sheetlike slabs of intrusive rock called *dikes* or as *sills* (intrusive rock that parallels the boundary between two rock layers). In this way, multiple invasive events build a framework that supports the accumulation of lava flows and volcanic rubble to heights greater than those of cones that do not have such a framework (fig. 3-2). Composite cones can continue growing until their slopes exceed the threshold of stability and then suddenly collapse, as happened at Mount St. Helens.

Fig. 3-3. Panum Crater, on the shores of Mono Lake, California, contains the northern-most rhyolite dome in a chain of volcanoes extending from near the town of Mammoth in Long Valley Caldera. Charged with gas and encountering groundwater, the initial explo-sive eruption at Panum Crater formed a broad, low ring of volcanic ash. As the magmatic gas was depleted and the groundwater was sealed off from the magma, the explosions ceased and the magma oozed out slowly as an extremely viscous lava to form a craggy rhyolite dome inside the ash ring. The dome spread in two directions to make the visible crease across its center. (Photo: Richard V. Fisher)

Domes

On the outside slopes of many composite cones or within their craters are domelike protrusions of lava. *Lava domes* result from the slow extrusion of highly viscous silica-rich magma. Most domes are rather small, but some are volcanoes in their own right and have volumes exceeding 25 cubic kilometers. Although they may end as rather slow-moving lavas, dome-producing erup-tions may start very explosively, forming reamed-out pits rimmed with pyro-clastic debris (fig. 3-3). The explosive activity wanes as the gas content and pressure within the magma decrease. With lower gas pressures, the magma extrudes slowly as viscous rhyolite lava that forms thick stubby flows or bul-bous domes.

As a lava dome grows, its margins slowly creep outward with steep cliff-like margins and a surface covered with rubble composed of sharp-edged glassy

Fig. 3-4. Following the collapse and northward explosive blast of Mount St. Helens on May 18, 1980, a 1,220-meter-deep, horseshoe-shaped crater remained. A silicic dome, now largely quiet, grew within the crater through the 1980s. The dome is still being monitored by the Cascades Volcano Observatory located at Vancouver, Washington. This photograph was taken in August 1982. (Photo: Richard V. Fisher)

blocks. If the protrusion occurs on a steep slope, dome margins can collapse in a dangerous mass of hot rubble that can form pyroclastic flows (see chap. 5). A dome has been growing slowly within the crater of Mount St. Helens since the eruption of 1980 (fig. 3-4). Domes have also filled the crater of Mount Pelée, Martinique, and of many other volcanoes.

Calderas

A *caldera* (from Spanish for "cauldron") is a very large crater formed when the ground surface collapses as a result of the extrusion of such large amounts of ash, pumice, and rock that the magma chamber is partially emptied; the ground above the magma chamber then caves into the momentary void. The dimensions of calderas are quite variable, from as small as a few kilometers to as large as 60 kilometers in diameter. The large ones are easily visible from space, but on the ground they are difficult to recognize because their configuration is not visible from a single viewing place. Eruptions of large volumes of material that create calderas usually give rise to hurricane-like clouds of hot volcanic debris that deposit ash and pumice over widespread areas. Chapter 5 is devoted to a discussion of these clouds.

Cinder Cones (Scoria Cones)

Cinder cones, or *scoria cones*, are relatively small volcanoes composed of basaltic fragments and are one of the most common volcanic landforms on Earth. (The term *cinder* is generally used by geologists in the United States, whereas the term *scoria* is used by geologists of the United Kingdom.) Cinders are small nut-size to fist-size, or larger, pieces of red or black lava containing abundant bubblelike cavities (called *vesicles* by geologists) that cool in the air after extrusion from a vent. Cinder cones are usually constructed completely of these fragmental cinders and form steep-sided mounds with a small crater at the top. They commonly occur in clusters, sometimes associated with more than a hundred neighboring cinder cones. They also occur on the slopes of other types of volcanoes, such as shield volcanoes, which are discussed below. Cinder cones are discussed further in chapter 4.

Maars

A *maar* is a small volcano with a wide crater ranging from several hundred meters to one or two kilometers in diameter. Their crater floors commonly lie below the general level of the surrounding topography. Maars usually form from steam explosions that occur when rising magma comes in contact with and mixes with groundwater or surface water. These volcanoes are discussed further in chapter 4.

Shield Volcanoes

As the name suggests, *shield volcanoes* are broad and have low slopes, much like an overturned shield. The big island of Hawaii is actually five coalesced shield volcanoes, Mauna Loa and Kilauea being the youngest. Of the other three—Kohala, Mauna Kea, and Hualalai—only Hualalai has erupted in historic time. The Hawaiian shield is so massive that Mauna Loa stands higher from base to top than any mountain on earth—9,090 meters from the floor of the ocean to its summit and is 600 kilometers in circumference at its base. Shield volcanoes are constructed of solidified basaltic lava that was originally in a highly fluid state, somewhat like hot syrup, and moving at speeds up to 12 kilometers per hour—exceptionally fast for lava. Such lavas can flow long distances and therefore construct gentle slopes and broad summit areas, unlike the steeper-sided composite volcanoes or cinder cones. The other Hawaiian islands, such as Oahu, Maui, and Kauai, are also basaltic shield volcanoes, but erosion has so greatly carved their original shapes that only the discerning eye of a volcanologist can identify them as such. The shield structure of many of the volcanoes that make up the Hawaiian Islands are also modified somewhat by younger and smaller volcanic cones that grew upon them.

Volcanic Eruptions

A volcanic eruption is the expulsion of gases, fragmental material, or molten lava—or all three—from beneath the ground through a vent into the atmosphere. Different kinds of eruptions leave characteristic deposits that build the different kinds of volcanoes discussed above.

How long do eruptions last? This seemingly straightforward question is complicated by the meaning of "volcanic eruption." One complication concerns the different kinds of volcanic activity—ranging from earthquakes caused by ascending magma to volcanic gases emitted, and from explosive ejection of volcanic ash to quiet effusion of lava. Suppose a volcano explodes once a day for three weeks, is quiet for three weeks, then renews activity for another three-week period and continues with the same pattern of activity for ten years before it stops. What do you call the eruption: Each explosion? Each three-week interval of activity? Or the entire ten-year episode?

In the vocabulary of some volcanologists, including the authors, a volcanic eruption is divided into (1) an eruptive pulse, for example, a single explosion that may last a few seconds to minutes; (2) an eruptive phase, in which strong explosions cause pulsating eruption columns that may last a few hours to days and consist of numerous eruptive pulses; and (3) a single eruption, composed of several phases, that may last a few days to months and, in some volcanoes, for years. If lava is emitted with, or instead of, explosive fragments, the same time frame is used to designate volcanic eruptions.

Volcanologists at the Smithsonian Institution use the quiet intervals between active extrusions and explosions to bracket a complete eruption. An eruption that follows its predecessor by less than three months is said to be a phase of the earlier eruption unless the type of volcanism is distinctly different. This definition creates the interesting (but unlikely) possibility that a newcomer to a region could buy property on an erupting volcano during a quiet three-month period without knowing that the volcano was in eruption.

Gas Eruptions

It was August 21, 1986, in the highlands of Cameroon, West Africa. Night had just fallen, and Lake Nyos, one of thirty small volcanic crater lakes distributed across Cameroon, was quiet (fig. 3-5). Hadari, a Fulini cattle-herder, lived with his family in their hillside home, from which they had a view of Lake Nyos, far

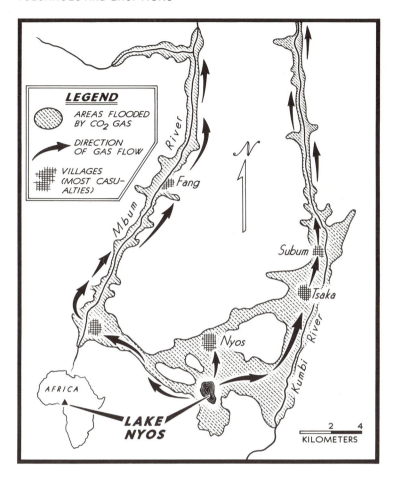

LEGEND

AREAS FLOODED BY CO_2 GAS

DIRECTION OF GAS FLOW

VILLAGES (MOST CASU-ALTIES)

River

Mbum

Fang

N

Subum

Tsaka

Nyos

Kumbi River

AFRICA

LAKE NYOS

2   4
KILOMETERS

Fig. 3-5. Lake Nyos is a very deep crater lake in the Cameroon Highlands of West Africa. On August 21, 1986, it erupted colorless, odorless, carbon dioxide gas. Being heavier than air, the carbon dioxide flowed from the volcano into the valleys below. It replaced the air in the valleys, suffocating all animal life within the areas shaded on the map. 1,700 people died in the villages below Nyos.

down the grassy slope. A mile below Lake Nyos in the valley was the village of Nyos. Suddenly, Hadari was awakened by loud rumbling. He and his family rushed from the house. In the darkness they saw a column of vapor rise from the lake and pour over the rim of the volcano and down the valley toward Nyos. Frightened, Hadari and his family hurried to higher ground and spent the night. The eruption spewed carbon dioxide gas, which, being heavier than air, moved like a river of smoke 50 meters thick and 16 kilometers long to engulf the town of Nyos and its vicinity, where twelve hundred people died. In the nearby villages of Subum and Cha, more than five hundred people died. In the morning light, three thousand cattle could be seen scattered around like litter, but they were untouched by flies or vultures, for the scavengers had also died. Death had come to all of the animals. People lay dead in doorways, beds, and beside their late-evening meals. The river of carbon dioxide had displaced normal air or mixed with it, and the people went to sleep forever in the silent flow—a concentration of only 15 percent carbon dioxide in air is toxic. This eruption was unusual, for it consisted only of gas.

But gases can be present without an eruption. It has been determined, for example, that carbon dioxide is seeping through the ground from Mammoth Mountain, a large, complex dome volcano on the edge of Long Valley Caldera, California, which is one of America's largest ski areas. The gas is seeping upward from active magma that is several kilometers below the surface. On March 11, 1990, Fred Richter, a forest ranger at Mammoth, was caught in a blizzard and sought refuge in an old cabin that showed itself above the snow drifts. He descended into the cabin through a hatch and suddenly experienced the sensation of being suffocated as his pulse began to race uncontrollably. He immediately went to check for fumes from the cabin's fireplace, and in his words related what must have been experienced by the people at Cameroon: "I never made it. I started to see stars and my legs became weak. It was like you'd run a mile in three minutes and you couldn't get enough air into your body" (Reich, "Likely Site," p. A3). Instead of lying down to rest, Richter struggled back up the ladder and outside, where he quickly revived in the fresh air.

Carbon dioxide can rise through the rocks of the volcano to the surface, and because it is a dense gas, it can collect in pockets protected from the wind. Particularly susceptible are forest cabins where winds cannot disperse the gas. On Mammoth Mountain, carbon dioxide has killed trees in at least four areas around the volcano, the largest area being twenty-eight acres. The trees are not dying from surface gaseous emissions, but rather from underground concentrations of gas that affect their root systems. People walking or skiing through the area are not in danger, but a person who stops to sleep in a small depression out of the wind or to camp in a low area overnight could suffocate. There is no other volcanic activity associated with Mammoth Mountain at this time.

Hawaiian Eruptions

Hawaiian eruptions commonly occur as gusherlike lava fountains that generate red-hot lava rivers. Some eruptions thunder out of the ground from elongate fissures or tubular vents as roaring jets of gas, carrying liquid granules and blobs to form fountains reaching high into the sky (fig. 3-6). The jetlike incandescent gases stream rapidly upward through the magma, tearing it apart into

Fig. 3-6. Spectacular curtains of incandescent lava that shoot from elongate narrow fissures can exist because basaltic magmas are highly fluid. Beneath the ground, high pressures keep gases contained, but near the surface, gases expand and are rapidly released where pressure is reduced. Depending upon the length and openness of the fissures, the series of lava fountains may be a few hundred meters to many kilometers long. This example is from a fissure eruption of Kilauea Volcano, Hawaii. (Photo: U.S. Geological Survey)

lapilli- and bomb-size liquid blobs that partially or completely congeal while flying through the air. The terms *lapilli* and *bomb* refer to size: lapilli are 2–64 millimeters; bombs are larger than 64 millimeters.

Strombolian Eruptions

Strombolian eruptions, named after Stromboli Volcano in Italy, produce high-arcing, incandescent "rooster-tails" reminiscent of colorful fireworks fountains. At night, the glowing, red, ballistic particles are spectacular. The ejecta are basaltic to andesitic cinders and bombs that rain down around a vent to construct cinder cones (see chap. 4).

Plinian Eruptions

Plinian eruptions produce ash columns that extend as high as 50 kilometers into the atmosphere (fig. 3-7), where high winds can spread ash for hundreds

Fig. 3-7. Hot gases and volcanic ash particles in explosive eruptions are carried tur-
bulently aloft, at first by a high-velocity "jet" of volcanic gas. As the heated upper part of
the eruption column incorporates air and becomes lighter, it rises buoyantly like a hot-air
balloon on a cold morning. This eruption column is from Mayon Volcano, Philippines,
during an eruption on April 26, 1968. (Photo: Arturo Alcarez, Philippine Commission on
Volcanology)

to thousands of kilometers from the volcanic vent. Depending upon the speed
of the stratospheric winds and the height of the eruption column, ash may be
carried around the world more than once from a single Plinian eruption. Ash
columns begin their evolution beneath the ground where dissolved gases expand within the magma as it rises into zones of lesser pressure. Starting at
shallow depths, 1–2 kilometers, the magma disintegrates and continues to the
surface to rise into the atmosphere as an eruption column (see chap. 5).

The name *Plinian* is from Pliny the Elder, a Roman nobleman who died during the A.D. 79 eruption of Vesuvius as he was trying to rescue people and was later immortalized by his nephew, Pliny the Younger. The younger Pliny witnessed the eruption when he was eighteen years old and wrote his recollections about A.D. 85. He likened the shape of the eruption column to that of a pine tree (the Italian Stone Pines have an umbrella-like top):

> A cloud, from which mountain was uncertain at this distance, was ascending, the form of which I cannot give you a more exact description of than by likening it to that of a pine tree, for it shot up to a great height in the form of a very tall trunk, which spread itself out at the top into a sort of branches; occasioned, I imagine, either by a sudden gust of air that impelled it, the force of which decreased as it advanced upwards, or the cloud itself being pressed back again by its own weight, expanded in the manner I have mentioned; it appeared sometimes bright and sometimes dark and spotted, according as it was either more or less impregnated with Earth and cinders. (quoted in Bullard, *Volcanoes*, pp. 138–39)

Two Eruption Types

Mount Pinatubo's Big Surprise

The eruption of Pinatubo Volcano in 1991 was one of the biggest in the twentieth century. Mount Pinatubo, a composite volcano in the Philippines, had last erupted four hundred years ago, but that eruption had long been forgotten, and the new one came as a huge surprise. It was a large Plinian eruption, producing abundant pyroclastic flows extending outward from its crater. Ash piled up, filling all of the valleys and depositing a vast store of loose volcanic ash. (Details of the eruption are further discussed in chapter 15 as a model for successful mitigation of volcanic hazards.) Each year since the eruption, the loose ash has mixed with the yearly monsoon rains and has created devastating mudflows (fig. 3-8).

In addition to causing destruction, volcanic eruptions can even affect global politics, and the fate of Clark Field is an example. The immense U.S. Air Force base, once the bulwark of American presence in the Pacific region, is dangerously close to Mount Pinatubo and has now been closed permanently and turned over to the Philippine government. Closing of the base also resulted in the loss of thousands of jobs for Philippine residents.

The fertile volcanic soil of the Central Plain, which is north of Manila on Luzon, in the Philippines, is that nation's richest agricultural region. With a good irrigation system and favorable weather, three rice harvests a year are possible. This fertile plain is adjacent to the composite cones and large lava domes of the north-south-trending Zambales Range (fig. 3-9). Clark Field, the largest U.S. air base outside of the United States, was constructed within the plain. The base began as a small cavalry post in 1902 but took on strategic importance in the South Pacific with the advent of military air power, and it

Fig. 3-8. The climactic phases of the Mount Pinatubo eruption occurred on June 12, 1991, with eruption plumes reaching an elevation of 25 kilometers. The gas-thrust portion of the eruption column is hidden in this view. The turbulent, convective portion is visible, rising on hot gases until the gases and contained ash particles have cooled to the same temperature as the surrounding air. (Photo: Karin Jackson, U.S. Geological Survey)

grew, usurping rich agricultural lands. It became the first Japanese target in the Philippines during World War II.

The deceptively quiet volcanic chain adjacent to the Central Plain extends from the Lingayen Gulf to Mount Bataan on Manila Bay. Mount Pinatubo was the highest volcano on this chain, rising to 1,745 meters above sea level before its 1991 eruption. Volcanoes of the Bataan Peninsula, at the southern end of the Zambales Range, are visible from Manila, except on days of dense smog. These mountains were home to the seminomadic Aeta people, who lived on the slopes, raising rootcrops and bananas. In 1941, American and Philippine prisoners of World War II were forced to walk the infamous Bataan Death March across the Central Plain.

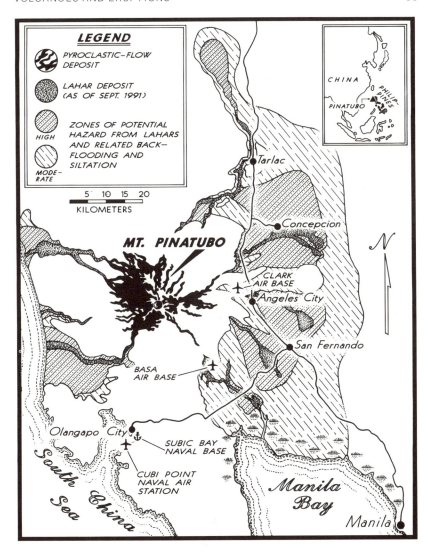

Fig. 3-9. Mount Pinatubo, central Luzon, Philippines, erupted violently in 1991. It is but one of many volcanoes that make up the north-south trending Zambales Range, which is lightly populated, rugged terrain. To the east of Mount Pinatubo is the rich farmland of the Central Plain and also Clark Air Base. To the west, bordering the South China Sea, is farmland, some of which was abandoned after the eruption. Pyroclastic flow deposits from the eruption filled deep valleys radiating from Pinatubo's summit, but the ash washed from the deposits caused (and is still causing) floods across the region. Lahars sent pyroclastic debris down valleys and well beyond the stream banks. Many farms and towns, as well as Clark Air Base, had to be abandoned after the eruption. (Based on a map prepared by the Philippine Institute of Volcanology and Seismology and the U.S. Geological Survey)

Fig. 3-10. The tranquillity of the Zambales Range was broken on April 2, 1991, when superheated steam created a line of small explosion craters across the summit of the Pinatubo lava dome. (Photo: Chris G. Newhall, U.S. Geological Survey)

The volcanic origin of the Zambales Range and possibility of any sort of volcanic eruption had not been considered by the farmers of western Luzon nor by the high-tech aviators who flew in and out of Clark Air Base. Because the eruptions are so infrequent, they did not know that the Zambales volcanoes can erupt destructively. No one had studied the volcano, for the dangers had been forgotten over the tens of generations of inactivity, and not even legends of activity had survived.

The only people aware of the potential volcanic activity of Mount Pinatubo were the geothermal exploration geologists of the Philippine National Oil Company and of Philippine Geothermal, Inc., who were searching for young volcanoes in order to tap these natural heat sources and generate electricity. They were looking for volcanoes that erupt infrequently and produce large volumes of volcanic ash. Such volcanoes leave large heat sources beneath the surface after an eruption because not all of the magma is ejected from the ground during volcanic activity, and what is left cools extremely slowly over hundreds of thousands of years. The heat sources generate hot water and steam (fig. 3-10), which are used to run electricity generators.

In their search for a geothermal system, the geologists demonstrated that in the past Mount Pinatubo had produced large amounts of ash as well as debris flows that extended as far as 20 kilometers from the mountain, and they determined the Pinatubo deposits to be between six hundred and eight thousand years old. They drilled three geothermal exploration wells on the volcano but then plugged them up because the fluids were too hot (as high as 338°C) and too acid to use for an electricity-generating plant.

Volcanoes that erupt infrequently, like Mount Pinatubo, are especially dangerous because their history is poorly known. With few or no legends, there is nothing to remind local people of the danger from the volcano. Such volcanoes, mostly located along plate margins, are far more hazardous than the long-lived basaltic volcanoes, like Kilauea and Mauna Loa, that constantly erupt small volumes of cinders and lava and keep people aware of the danger.

Hawaiian Eruptions

Hawaiian eruptions, which are generally of basaltic magma, destroy property, but the lava flows produced by them rarely catch people off guard. Plenty of warning is given, and most lava flows are slow enough to be outrun and avoided. Not all of the five million tourists who visit Hawaii each year want to visit vast barren lava fields or smell sulfur gases, but many of them make an effort to see a volcano in eruption. On the dry west coast (Kona Coast) of the island of Hawaii, which is locally known as the Big Island, and in the island's main city of Hilo, visitors can catch helicopter or plane flights to view the eruptions.

Even without an eruption, Hawaii, the newest island of the chain, is an exciting adventure. The remains of recent volcanic action are everywhere: steaming craters, fumaroles, fresh *pahoehoe* (smooth) and *aa* (rough) lava flows. The ghosts of High Chief Keoua's warriors are preserved in their footprints—all that is left of a party of warriors that was overwhelmed by an eruption from Halemaumau in 1740 while they were stalking the army of Kamehameha. This incident provided proof to many Hawaiians that Madame Pele, goddess of the volcanoes (see chap. 10), gave Kamehameha special protection, and so the Hawaiians made him the chief ruler of Hawaii.

Erosion has carved the other islands of Hawaii into hills, valleys, and mountain peaks that no longer look like volcanoes. But the volcanic heritage can be seen in lava flows that are thousands of years old and that preserve the surface texture of their original molten condition—some with smooth ropy pahoehoe, some with rough, velcro-like aa lava, which cuts the leather soles from boots and shoes, exactly like modern aa lavas. Most features of shield volcanoes and their lavas are found in the easily accessible Hawaii National Park, which encompasses the broad summit area of Kilauea volcano and its active pit crater, Halemaumau (fig. 3-11).

Many Hawaiian eruptions occur when streaming gases from subterranean magma chambers carry blobs of yellow-and-red-hot plastic lava to form fountains that may extend 1,000 meters or more into the sky (see fig. 3-6). Some lava fountains build spatter cones as the semi-molten globules accumulate

Fig. 3-11. Looking northeast across the summit of Kilauea Volcano, it is easy to see the overlapping collapse pit craters that mark the summit. These craters formed by pistonlike collapse of the summit over a near-summit magma chamber. Magma drained into subterranean fissures leading into the flanks of the volcano where it erupted from fissures at a lower elevation. (Photo: U.S. Geological Survey)

around a vent. In some cases, the lumps from a fountain remain hot and molten and coalesce into liquid lava, becoming feeders for large lava flows. Continuously fed from vents at the summit or along rifts that break the volcano's flanks, red-hot lava moves downward in pulsating streams of irregular width, carrying islands of black crust that float downstream or tip up and descend into the depths of the flow. Hour after hour, day after day, lava fountains may pulse upward hundreds of meters and then subside, only to pulse upward again and feed the lava flows that course down the sides of the volcano toward the sea. Millions of cubic meters of lava that reach the sea build the shoreline outward and enlarge the island—new land in a blue ocean to eventually provide nourishment for terrestrial plants and animals.

Mauna Loa is barely over a million years old, which is geologically young, yet its volume is a colossal 40,000 cubic kilometers. Cliff erosion in streams and exposures in craters reveal thousands of solidified lava flows stacked one above the other like pages in a book, proving that a single cataclysmic eruption did not make this vast mountain. Instead it was slowly built by steady, frequent outpourings of a few thousand to a few hundred million cubic meters of lava at a time (one cubic kilometer is equal to a billion cubic meters). Since 1843, 13 percent of Mauna Loa's surface has experienced a lava flow, but historically, average eruption rates change. From 1846 to 1876, the average was 45 million cubic meters per year. Since then, the rate has decreased by half.

Because explosive eruptions are infrequent on Hawaii and consist mostly of lava flows, the risk to human life from eruptions is very low. Throughout the last two hundred years, the eruption rates of Mauna Loa and Kilauea have been about one eruption every two or three years. At the current average eruption rate of about 22 square kilometers per year, the surface of Mauna Loa will be coated with 2 meters of lava in four thousand years. Most lava flows from Mauna Loa occur only in unpopulated areas, but along the south Kona Coast are several new housing developments, which in some cases are situated on lava flows only a few decades old. Particularly widespread lava flows inundated this area in 1887, 1907, 1919, 1926, and 1950. Residents living on these historic lava flows along the Kona Coast need not worry that flows will return soon to the same areas, but planning a home to last for several generations may end in disappointment (fig. 3-12).

Kilauea Volcano now erupts more often than Mauna Loa and has destroyed many works of man, as it did the village of Kapoho and the surrounding farmland in 1959–1960. In December 1965, the ground around Kapoho was still warm. And from February 1972 to July 1974, Mauna Ulu ("growing mountain"), on Kilauea's east rift, erupted continuously for 901 days, producing 162 million cubic meters of lava that covered 46 square kilometers. Lava reached the seashore through a 12-kilometer-long system of lava tubes and entered the sea, producing great quantities of steam with much hissing, sometimes explosively as a wave broke over a lava tube and trapped the steam inside. Sometimes, the red-hot lava slipped into the sea with surprisingly little fanfare, but when confined in small spaces, the water and lava produced small explosions. Without such confinement, the overwhelming volume of ocean water simply couldn't be heated fast enough by the comparatively small volume of lava to cause an explosion. Lava flows from this particular eruption severed large sections of the Chain of Craters Road that linked the national park to the southeast coast, but there was little loss of private property, for the lava from this eruption stayed within national park boundaries. The Mauna Ulu eruption was particularly educational for the curious public, and therefore beneficial to the tourist industry, because access was possible to Mauna Ulu via the Chain of Craters Road. Lava lakes and minor lava fountaining were visible from safe observation points established by the National Park Service and the Volcano Observatory.

On January 3, 1983, Pu'u O'o ("hill of the O'o bird") began to erupt along Kilauea's east rift. It still continues at the time of this writing (July 1996) and

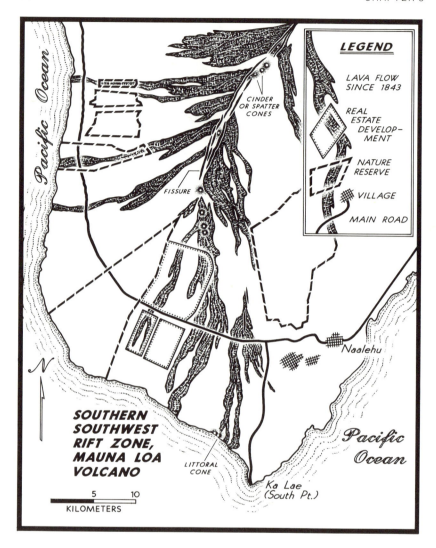

LEGEND

LAVA FLOW
SINCE 1843

REAL
ESTATE
DEVELOP-
MENT

NATURE
RESERVE

VILLAGE

MAIN ROAD

Pacific Ocean

CINDER
OR SPATTER
CONES

FISSURE

Naalehu

Pacific
Ocean

N

SOUTHERN
SOUTHWEST
RIFT ZONE,
MAUNA LOA
VOLCANO

LITTORAL
CONE

Ka Lae
(South Pt.)

5   10
KILOMETERS

Fig. 3-12. The conflict between volcano and cultural developments is evident along the southwestern rift zone of Mauna Loa Volcano, Hawaii, where lava flows that have erupted since 1843 are now part of several large real estate developments. Future eruptions could endanger residents, and most flows certainly will cut the main highway.

Fig. 3-13. A common occurrence on Hawaii—a highway being crossed, buried, and burned by a lava flow. This one is near what was once the village of Kalapana, which was buried and bulldozed by lavas that flowed several kilometers from the east rift of Kilauea Volcano. The highway department usually waits for an eruption to end before it takes action. If the flows are thin, the department bulldozes a new road across the flows. (Photo: Grant Heiken)

has destroyed habitable land and villages along Hawaii's east coast. It is the most voluminous rift zone eruption of the past two centuries, its flows having added 1.5 square kilometers of new land to the island. The eruption started with a spectacular fountain of lava along the central east rift, which then closed in on the edges and became a single vent that built a hill 255 meters high made of lava-fountain spatter. Lava flowed through lava tubes to the coastline, where it engulfed 181 homes, several businesses, a national park visitor center, the village of Kalapana, and a well-known black sand beach at Kaimu. Since the eruption began, the volume of lava emitted has been about 10 million cubic meters per month. By 1993, lava had covered an area of 83 square kilometers, and the geologists who counted eruptive episodes had recorded fifty-three distinct surges of activity (fig. 3-13).

As at Mauna Ulu, lava flowing into the sea creates a drama that transports the viewer back to what might be a glimpse of primitive earth—hissing, sometimes deafening, plumes of steam intermingling with incandescent streams of lava issuing from tubes opening on the shore and beneath the water.

Explosions occasionally echo from within the thundering din as waves break across the uneven rocky lava delta and water swashes into the tubes. Barren basalt lava forms a black desert without color, without a single growing sprig. A tourist entwined with camera straps remarks, "This has to be a preview of hell," while the thunk, thunk, thunk of helicopters laden with tourists pace back and forth in the sky. Most tourists are held back from hell at approved viewing sites by diligent park rangers.

Slowly moving, red-hot lava bulldozes houses and destroys fertile land in dramatic episodes of devastation, but the Mauna Loa and Kilauea eruptions also indirectly debilitate the more distant residents of the Big Island. Basaltic lava with its high sulfur content releases sulfur dioxide gas, which when released into the atmosphere and mixed with steam, converts to very small aerosol droplets of sulfuric acid (see chap. 9). A blue, sulfuric acid aerosol at times forms a haze across the island, obscures the Hawaiian landscape, and causes discomfort for people with respiratory ailments. Even normal drinking water from wells can become contaminated as acid rainwater seeps through the lava, following fractures into the water table. Drafts of water from ordinary water wells are not possible to drink, so residents use catchment areas (roofs or special rain sheds) to collect rain water in cisterns. But even collected rainwater can sometimes be polluted by airborne sulfuric acid.

Many visitors are fatalistic about Hawaii's volcanoes and declare the frequent eruptions and lava flows to be the will of Madam Pele, the beautiful but unpredictable red-headed goddess of volcanoes. The many legends of Pele are oral traditions that recall past volcanic activity and establish an awareness of future activity—an awareness justified by scientific studies of the thousands of lava flows that make up the exotic Big Island.

THE "wonder fluid" of our planet is water. Plants and animals not only require water for life, they could not have evolved without it. Living cells are mostly water. It is close to being a universal solvent and is one of the few natural fluids in the universe which, upon freezing to a solid (ice), actually expands, becomes less dense, and therefore floats upon itself. If that were not so, ice formed in the oceans would sink and the oceans would eventually freeze solid; life could not have evolved without liquid water. As we have seen, water also affects volcanoes. It can transform ordinary volcanic eruptions into steam eruptions and is responsible for the development of maar volcanoes in lieu of cinder cones. We will examine the influence of water on eruptions next.

References

Bullard F. M. *Volcanoes*. Austin: University of Texas Press, 1968.
Crandell, D. R., and D. R. Mullineaux. *Potential Hazards from Future Eruptions of Mount St. Helens Volcano, Washington*. U.S. Geological Survey Bulletin 1383-C. Washington, D.C., 1978, pp. 1–26.

Crandell, D. R., D. R. Mullineaux, and C. D. Miller. "Volcanic-hazards studies in the Cascade Range of the Western United States." In *Volcanic Activity and Human Ecology*, edited by P. D. Sheets and D. K. Grayson, 195–219. London: Academic Press, 1979.

Freeth, S. J. "The anecdotal evidence, did it help or hinder investigation of the Lake Nyos gas disaster?" *Journal of Volcanology and Geothermal Research* 42 (1990): 373–80.

Lipman, P. W., and D. R. Mullineaux, eds. *The 1980 Eruptions of Mount St. Helens.* U.S. Geological Survey Professional Paper 1250. Washington, D.C., 1981.

Reich, K. "Likely site of eruption at Mammoth reassessed." *Los Angeles Times*, November 27, 1994, p. A3.

Simkin, T. and L. Siebert. *Volcanoes of the World.* 2d ed. Tucson, Ariz.: Geoscience Press, 1994.

Stager, Curt. "Silent death from Cameroon's killer lake." *National Geographic*, September 1987, 404–20.

Westervelt, W. D. *Hawaiian Legends of Volcanoes.* Rutland, Vt.: Charles E. Tuttle, 1979.

Wilcox, R. E. *Some Effects of Recent Volcanic Ash Falls with Especial Reference to Alaska.* U.S. Geological Survey Bulletin 1028-N. Washington, D.C., 1965, pp. 409–76.

———. "Volcanic ash chronology." In *The Quaternary of the United States*, edited by H. E. Wright, Jr. and D. G. Frey, 807–816. Princeton: Princeton University Press, 1965.

Poseidon and Pluto:
Water and Volcanoes

Frontispiece. Like water draining from a bathtub, this dramatic maelstrom of water is spiraling down the throat of Capelinhos Volcano during a lull in activity. But in this case, the water descends to meet hot magma to cause renewed phreatic (steam) explosions. The eruption occurred in shallow water off the coast of Faial Island, Azores, in 1957–1958. (Photo: T. Pacheco, Serviços Geológicos de Portugal)

In 1831, Charles Darwin, then twenty-two years old, set out on a five-year adventure as the naturalist on HMS *Beagle*, his only goal being to explore the physical and biological world wherever the *Beagle* traveled. Darwin's innate powers of observation and deduction led to his revolutionary accomplishments in the field of biology. He was also an accomplished geologist and made many geological discoveries, which are reported in his 1844 book, *Geological Observations on Volcanic Islands*. He was the first to describe the effects of seawater on the formation of volcanoes and explained the origin of coral atolls as being the tops of volcanoes that have sunk into the sea.

HMS *Beagle* made landfall at several islands in the Galapagos Islands, giving young Darwin an opportunity to describe the flora, fauna, and rocks. He described several small volcanoes that stood as islands in the sea. Their slopes were composed of layers of angular-to-rounded, finely broken basalt-glass fragments altered over time to a yellow-brown waxy or resinous substance. From these features, and because the volcanoes stood in water, Darwin concluded that they were "communicating with the sea" (Darwin, *Geological Observations*, p. 127). He noted that "when seen in mass its stratification, and the numerous layers of fragments of basalt, both angular and rounded, at once render its subaqueous origin evident. . . . The position near the coast of all the craters composed of this kind of tuff or peperino, and their breached condition, renders it probable that they were all formed when standing immersed in the sea." (ibid., p. 112)

The mention of volcanoes invokes visions of molten rock rising from beneath the earth's surface to explode upward into the atmosphere as high eruption columns or to pour onto the surrounding land as lava flows. But unlike Darwin, few people today are aware that water is a powerful influence on the type of eruption and the form of some volcanoes. There is a constant duel between Poseidon, mythical lord of the sea, and Pluto, mythical lord of the fiery underground. Since the 1960s it has been increasingly recognized that the interaction of magma and water on or near the earth's surface produces volcanic eruptions that differ from eruptions in which magma does not interact extensively with surface water. Such interaction can take place in shallow water in lakes, on the edges of islands, or on the deep ocean floor. In some cases where there is no surface water, magma encounters underground water as it rises, resulting in eruption of magma-water mixtures onto dry land.

Fig. 4-1. In this aerial view of Honolulu, the bowl-like shape of Diamond Head is typical of tuff rings built from violent, shallow steam explosions. Bathers on Waikiki Beach, familiar with the profile of Diamond Head, never suspect that it is the rim of a crater that houses government facilities. (Photo: Courtesy of Peter Mouginis-Mark)

Eruptions resulting from magma-water interactions are appropriately called *hydrovolcanic eruptions.*

When magma mixes with water at shallow depths or at the earth's surface where pressures are nearly atmospheric, it can explode. These explosions can excavate wide, shallow bowl-like craters and spread volcanic ash across broad, low-standing rims around the crater. Such volcanoes are called *maars,* from an old German word meaning "lake." In the East and West Eifel districts of Germany, there are many volcanoes with bowl-shaped craters that contain lakes filled by rain. All of these are maars (fig. 4-1). Without the water-rich environment, cinder cones, rather than maar volcanoes, would be formed, as discussed below.

The greatest and most powerful hydrovolcanic explosions occur when all of the water available to a fixed amount of magma is transformed into steam. Research indicates that this happens when the proportion of water to magma is about three parts water and one part magma, but investigations are still continuing. Steam explosions caused by surface water mixing with magma commonly reduce solidified magma to much smaller fragments than do explosions caused by gas expansion from within the magma. Strombolian explosions produce molten or plastic clumps that solidify and construct cinder cones, but when the magma mixes with water, steam explosions can convert the magma to small ash particles with the texture of fine sand or flour and make craters that form maar volcanoes. It is not surprising, then, that maars are commonly constructed in watery environments—in lakes, along seashores, or on land that lies over abundant underground water. History's most recently publicized hydrovolcanic eruptions were at Ukinrek, Alaska, in 1977, which made a small bowl-like crater with hydrovolcanic deposits draped over its rim.

The maar volcanoes described by Darwin in the Galapagos erupted within the sea. This kind of volcano also forms on dry land, as happened in Death Valley (Ubehebe Crater) and in many other places in the American West where the land is underlain by large reservoirs of underground water. Within many maars, some strata are wavy rather than flat, shaped somewhat like rippled sand dunes, and at one time were thought to originate from strong atmospheric winds during eruptions. This notion was changed by two modern events: a nuclear explosion in the South Pacific and the eruption of Taal Volcano, Philippines, in 1965.

Bikini and the Base Surge

Immediately following World War II, the United States continued experiments with nuclear bombs. In July 1946, the U.S. Army and Navy detonated two nuclear bombs in the lagoon of Bikini Atoll just north of the equator in the Pacific Ocean—one above water and one under water—to determine what would happen to a naval fleet under nuclear attack. Anchored in the lagoon were expendable ships of the U.S. Navy and ships of the captured German and Japanese fleets, including battleships, cruisers, an aircraft carrier, and submarines. Both bombs damaged and sank ships, but in different ways. The bomb that was dropped from an airplane and detonated above water burned and crushed some ships, and some of them sank; the underwater blast created an explosion column made of water droplets and gases mixed as an aerosol that swamped and sank some ships and formed a hurricane-velocity, horizontally moving cloud called a *base surge* (fig. 4-2). The explosion column, being heavier than air, collapsed back to the surface of the lagoon and surged laterally outward from the base of the explosion column, hence the name.*

In hindsight, it was an ill-conceived experiment with respect to the earth's environment, for we now know that it released harmful radioactive substances into the atmosphere. But from a volcanological point of view, considerable knowledge was gained because in subsequent years the discovery of the base surge at Bikini had a great impact on research concerning maar volcanoes and the effects of mixing water with magma. These avenues of research led to a better understanding of volcanic hazards, which in turn has helped save people living near erupting volcanoes.

* The underwater test proved to be most damaging because the base surge mist, like a pervasive fog, thoroughly drenched all of the ships with radioactive substances. The code name for the atom bomb tests was Operation Crossroads, with its human side chronicled in detail by Jonathan Weisgall, a lawyer who has represented the displaced natives of Bikini Atoll since 1975.

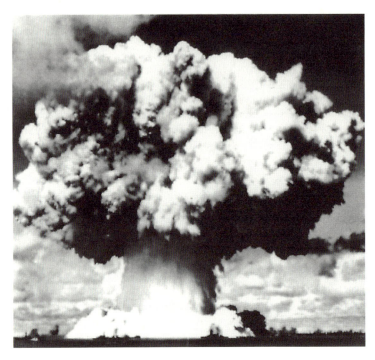

Fig. 4-2. An atom bomb detonated under 90 meters of water on the floor of Bikini lagoon beneath a naval fleet of dispensible World War II ships (Japanese, German, American) created a powerful explosion column. The width of the column is a little less than a kilometer. Instead of continuing to rise, the column collapsed downward to the water surface and generated a surge that swamped many ships as it moved away from the base of the column. (Photo: From Glasstone, 1950)

Additional information about base surges came sixteen years after the Bikini experiment, in 1962, when the U.S. government exploded a 100-kiloton atomic bomb called Sedan at a depth of 194 meters in alluvial deposits at the Nevada test site. The explosion excavated a man-made crater 370 meters in diameter and 98 meters deep. The fallback of its explosion column produced a base surge with an initial velocity of more than 50 meters per second as it spread across the desert floor. The Sedan base surge was a mixture of gases from the atomic explosion mixed with dust and sand from the desert floor.

The Bikini and Sedan tests illustrated the process by which an explosion column collapsed to produce a turbulent cloud that flowed across the surface of water and land. The Sedan base surge was formed by fallback of the explosion column that spread across the desert floor. This base surge, with its load of solid particles, deposited a layer of sand that had dunelike forms, which originated by the force of the winds generated by the explosion. This process was reproduced by Taal Volcano in the Philippines during its 1965–1966 eruption, in which magma mixed with water to produce a steam explosion. These three events, showing the fallback of an explosion column and the consequent development of base surges, led to a significant advancement in understanding how pyroclastic flows originate and to the recognition of their depositional features (see chap. 5).

Taal Volcano

A Philippine farmer, Severino de Ocampo, was living with his family on the southwestern shore of Volcano Island (now called Taal Volcano) in Lake Taal, Philippines, 60 kilometers south of Manila. He was awakened at about 2:00 A.M. on September 28, 1965, by his restless cattle. When he got up to investigate, he heard a loud explosion and out his window saw incandescent material shooting skyward like a large fireworks fountain.

At the same time, Contrado Andal was awakened. Contrado was the Volcanology Observer with the Philippine Volcanological Commission, which operated the volcanological station at Barrio Alas-as on the west-central coast of Volcano Island. The station was located about 800 meters west of what was then the vent. Before going to bed at 11 P.M. he saw no record of tremors on the seismograph, but when he was awakened later, there was a record of continuous vibrations, and outside the station, incandescent materials were being ejected from the volcano toward the southwest at about a 45-degree angle. Because of the direction of eruption, he advised his excited neighbors to evacuate the island immediately and to head north, advice that saved their lives. His boat normally held six, but he took twenty people and managed to save them all. Contrado went back to the station for gasoline an hour or so later but was kept away by a dense eruption cloud. The station was still intact at that time, but it later became buried under three meters of volcanic ash, sand, and boulders (fig. 4–3).

The eruption began as an ordinary Strombolian eruption with an incandescent lava fountain that would have normally constructed a cinder cone. But a crack later opened and let lake water pour into the vent to combine with the magma. This occurrence radically changed the behavior of the eruption from Strombolian to hydrovolcanic and produced base surges somewhat like the one at Bikini. But unlike that one, the Taal base surges were full of ash and larger fragments of volcanic rock. They swept over the landscape like hot hurricanes, tearing trees from the ground and sandblasting the trunks of those left standing. They flowed across the water and engulfed fleeing boats. About 190 people were killed by the Taal base surges.

As the base surges moved, fragments large and small rained down from the turbulent mass, leaving a carpet of deposits. And for the first time it was seen

Fig. 4.3

Fig. 4.4

that a volcanically produced, horizontally moving turbulent cloud composed of abundant particles piled up dunes that are more streamlined than desert sand dunes (fig. 4-4). A big difference between base-surge dunes and desert dunes is the presence in the first type of fragments too large to have been carried by ordinary winds.

Maar Volcanoes and Cinder Cones

Maar volcanoes are low-standing volcanoes with very wide, bowl-shaped craters, whereas cinder cones build high mounds with small craters at their tops (fig. 4-5). The difference between them stems from the fact that when basaltic magma mixes with water, the magma is blasted into fine-grained particles, and the resulting steam explosions excavate large craters to make maars. In the absence of abundant water, cinder cones or lava flows are formed. An example is located on the south side of the island of Oahu. Punchbowl, a maar volcano, its crater now a military cemetery, is close to the sea, but inland and at higher elevation farther from the coast are three cinder cones—Tantalus, Sugar Loaf, and Round Top. Although Punchbowl maar and the cinder cones have the same basaltic composition, the one difference is that Punchbowl, when it was an active volcano, had access to shallow seawater near the shoreline, whereas the cinder cones higher on ridges did not (fig. 4-6).

Cinder cones in eruption are small enough to watch in relative safety from a short distance away. They grow from thousands of recurring incandescent jets made of molten-to-plastic lumps of magma propelled by gases. The plastic lumps cool to solids with different shapes, often rough-surfaced, bubble-rich (vesicle-rich) fragments called cinders or scoria. The name *cinder* comes from their similarities to industrial slag and cinders produced by foundries.

One of nature's most spectacular shows occurs after sunset over an erupting cinder cone. Incandescent, luminous sprays of molten red lava light the night

---

Fig. 4-3. During the 1965 eruption of Taal Volcano, Philippines, water interacting with magma created a damped-down eruption column and a base surge that swept outward with hurricane force. In this 1965 photograph, the observers watch an expanding base surge with a 4-kilometer diameter at the time the photograph was snapped. The four men were at risk—the base surge could have expanded and moved across the water to overwhelm them. (Photo: Courtesy of the *Manila Times*, Manila)

Fig. 4-4. Debris carried across the ground by base surges is swept into dunes somewhat like sand dunes built by strong winds, as was this dune formed from the base surge at Taal in 1965, 600 meters from the volcano. Where eruptions have gone unobserved, such dunes give evidence that base surge eruptions have occurred. (Photo: Richard V. Fisher)

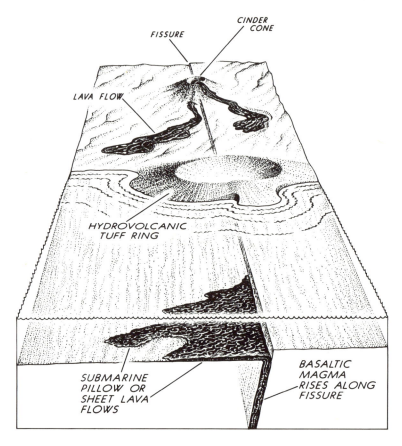

Fig. 4-5. Water influences the type of volcanic feature that is constructed. Inland, basalt rises to form lava flows or fountains in Strombolian eruptions to produce cinder cones. Near the water's edge, magma mixes with water to create violent, shallow, steam explosions that create wide bowl-shaped maar volcanoes. In deep water, pressure is great enough to dampen out explosions, and pillow or sheet lavas flow quietly across the seafloor.

sky like Roman candles on the Fourth of July. The lava droplets and clumps, formed by gases streaming through the magma and breaking it apart, spray outward in wide graceful arcs and hit the slopes as hot, already solidified or partially solidified cinders. Cinders continue to pile up around the vent to form a cone, with the slope's steepness limited by the stability of the loose cinders, much like a pile of dry sand. At times, the outer slope may get too steep, and masses of cinders and ash slump downward leaving a landslide scar that is soon covered over by the continued fallout of cinders. The higher the volcano

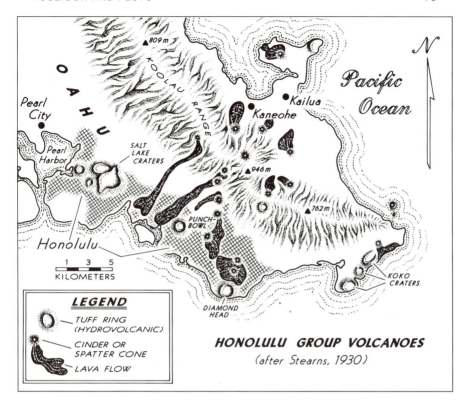

Fig. 4-6. The west end of Oahu, Hawaii, is a series of small, approximately 10,000-year-old volcanoes that erupted on both sides of the older volcano that forms the Koolau Range. Eruptions close to sea level formed maar volcanoes known as tuff rings, such as Salt Lake Craters, near the Honolulu International Airport; Punch Bowl, within which is the National Memorial Cemetery of the Pacific; Diamond Head; and Koko Craters. Eruptions at higher elevations built cinder cones and produced lava flows. This distribution of cinder cones and maar volcanoes indicates that interactions between magma and water are a controlling factor in the development of maars. (After Stearns, 1930)

gets, the slower it grows in height, for to grow in height and maintain a constant slope, its base must continue to grow outward in an ever widening circle that requires an ever increasing volume of cinders.

Poseidon and Pluto not only meet in shallow lakes and in water-saturated ground, but also at the edges and at the bottom of the sea, as well as under glaciers, where the heat of a volcanic eruption can melt the ice. Lava flows that encounter water when they enter the sea explode and pile up volcano-like mounds, *littoral cones*, on the shoreline. One well-developed littoral cone in Hawaii, Puu Hou ("New Hill"), was built in five days during a flow of lava into

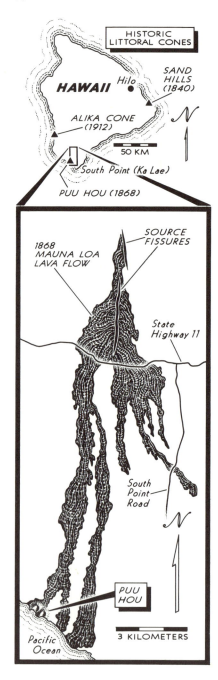

Fig. 4-7A. Littoral cones in Hawaii. There are three historic littoral cones in Hawaii. The Puu Hou littoral cone was produced when the Kahuku fissure eruption of 1868 from the south side of Mauna Loa flowed into the sea just west of South Point, Hawaii. Where lava met the sea, five days of continuous explosions built mounds of blocks and ash known as littoral cones.

Fig. 4-7B. Puu Hou littoral cone with Kahuku lava flow on both sides. Littoral cones look like volcanoes but have no vents that extend into the ground. (Photo: Richard V. Fisher)

the sea in 1868. It is easily visible from Ka Lae (South Point), the southernmost point of land of the United States. It looks deceptively like a small volcano but is rootless because it had no underground source and was fed only by lava that flowed from the land into water (fig. 4-7).

A Little-Known Frontier:
Volcanic Eruptions under the Sea

Imagine a science-fiction scene in which an undersea vehicle glides beneath the ocean's waters to monitor a reported underwater eruption. The observers at the window of the craft cannot see far through the darkened waters, even with powerful lights, and do not realize that an eruption plume is rapidly expanding beneath them, but instruments detect an unknown, lighter-than-water mass approaching the ship. Mystified but cautious, the captain decides to retreat and abandon the mission. The decision saves their lives because the lower-density

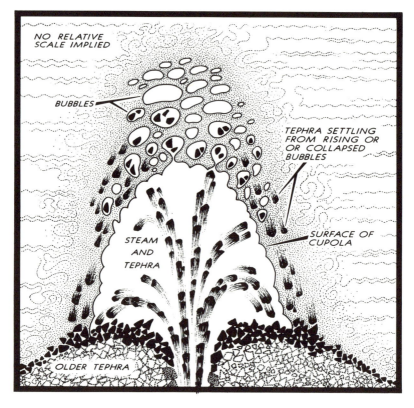

Fig. 4-8. Seawater is displaced by steam that makes a cupola enclosing an eruption jet. Spatter and clumps of lava erupted into the cupola do not contact seawater and fall back into the vent or on the crater rim inside the steam cupola.

mass is a bubble of expanding gases erupted from the underwater vent. Had it engulfed them, their craft would have entered a bubble of steam and plummeted into the erupting crater.

Science fiction? Maybe. Diving expeditions to the top of Surtla, an underwater volcano in Iceland, by Peter Kokelaar of the University of Liverpool, discovered lava bombs identical to those on land that are shaped by the drag forces of the atmosphere. Kokelaar concluded that the lava bombs that he found at the summit of Surtla, which are turned red by reaction with oxygen just like atmospheric lava bombs, were formed while traveling through an expanding gaseous underwater eruption cupola (fig. 4-8). They were sticky when they landed, as shown by adhered particles, so it is not likely that they had been ejected into or had settled through the water.

Fig. 4-9. Capelinhos Volcano, with one side open to the sea, near the coast of the Island of Faial, Azores, in early stages of growth in 1957. The island volcano later grew to connect with Faial. (Photo: Jovial, Serviços Geológicos de Portugal)

Where water is deeper than about 1 or 2 kilometers, explosions cannot normally occur because the pressure of the water keeps the volcanic gases from expanding quickly. Consequently, most eruptions on the seafloor give rise to lava flows. But shallow submarine volcanoes can be explosive enough to send columns of steam, water, pumice, and ash to the water surface and into the atmosphere. One such undersea eruption started at Faial which is off the coast of Capelinhos, Azores, on September 27, 1957, after four days of small earthquakes, when discolored bubbling patches surfaced on the sea about 0.8 kilometers from the coast. Two days later, before the volcano appeared above the water, steam and ash columns began to rise higher than 930 meters above the sea. In less than a month, a volcanic island rose to 140 meters (fig. 4-9). The island was eroded away by the sea and disappeared on October 25, but some days later the eruption became highly explosive (some columns rose to 7,750 meters) and soon formed a new island, which joined with the Capelinhos coast within a short time. Lava flows appeared in mid-December. The explosive activity ceased on October 25, 1958, and by that time had enlarged the area of Faial by more than 2.59 square kilometers.

An eruption of the undersea volcano, Myojin (Japanese for "bright god"), located 400 kilometers south of Tokyo, Japan, in the Bonin Islands chain, led to more serious consequences. The eruption began in September 1952 and was first detected at Point Sur and Punta Arenas on the coast of California where the U.S. Navy maintained two underwater listening stations called *sofar* (sound fixing and ranging). Recording stations listened in via cables running down the continental slope to hydrophones on the deep-sea floor. Sofar, designed to locate ships in distress as well as sounds of natural origin, detected Myojin's first explosion in the early morning hours of September 16. The sound waves, traveling nearly a mile per second, took ninety-seven minutes to reach the California coast. Twelve hours later a large blast sent a tsunami that traveled across the sea at 240 kilometers an hour and then struck Hachijo Island, 120 kilometers to the north. Soundings revealed that Myojin is the central vent of an 11-kilometer-wide caldera crater at the summit of an undersea volcano. A small volcanic island appeared at that spot and rose to about 30 meters above sea level as ash, steam, and lava globules continued to be ejected, building up the small island. The volcano and the sea competed in determining the height of the volcano. The sea frequently won and eroded the island away, but it would reappear, as the extruded projectiles increased in volume. The new islet was first seen on September 17 by the crew of the *Myojin Maru* and was still visible when visited by geologists on September 22.

The sea was calm on September 23 when the research vessel *Shinyo Maru* arrived. As they approached the site, there was one eruption when they were 8 kilometers away, and a second one when they were only 1.6 kilometers away. Professor Niino of the Tokyo College of Fisheries asked the crew to go forward to pick up floating pumice for a sample of the magma. As he lowered his net over the rail, the sea welled up several hundred meters away. Professor Niino described what followed:

> In the next moment, the swell looked like a boil on the surface. Then it bloomed, like a monstrous flower, 20 feet high and 600 feet across. Instantly water began to fall like a cataract while the dome's center rose to a height of 100 feet.
>
> Suddenly a black mass burst out from the dome's right side. Volcanic bombs sailed into the sky. Steam from the lower part of the cloud swirled across the sea to meet the boat. Hot pumice fell like rain, hissing as it dropped into the water. (quoted in Diet, "Myojin Island," p. 120)

The concussions pounded the crew while the *Shinyo Maru* retreated at full speed. As they sped away, a wave like a tidal bore caught the ship, tossing it as if it were a raft in river rapids.

On the same day, the *Kaiyo Maru V*, a ship of the Maritime Safety Board, was dispatched with scientists to study the activity. When they arrived on September 24, the island had disappeared. Apparently, the *Kaiyo Maru* had ventured to a spot directly above the crater, for at 12:21 P.M. Myojin erupted. All that was found was floating wreckage of the ship, some pieces with embedded volcanic bombs. The ship and its crew of thirty-one had been destroyed by a submarine volcanic eruption.

The response of lava to water depends greatly on water depth. In deep water below about 1 kilometer, such as on midocean ridges and deepwater sea-mounts, explosions rarely occur because the water pressure is greater than explosion pressures. Depending upon internal pressure of the lava, explosions may begin occurring at depths shallower than about 500 meters. Man's activi-ties at the ocean surface, however, are rarely affected by submarine eruptions. There is no need to worry about being engulfed in a volcanic eruption during your next ocean cruise.

Deep-Sea Eruptions

Exploration of the seafloors has revealed that there are more volcanic eruptions on the dark bottom of the sea than on land. Seventy percent of the earth's surface is below sea level, and all of that portion is volcanic terrain. Oceanog-raphers have mapped, sampled, measured, modeled, and drilled the ocean floor to come up with a fairly complete listing of undersea volcanoes and their composition, but there is still much to learn.

On the floor of the Pacific Ocean, it is estimated that there are 4,000 vol-canoes per million square kilometers; of these 4,000 volcanoes, 200 are over 1 kilometer in height (for comparison, Mount St. Helens is about 1.5 kilome-ters high). If these estimates can be extrapolated to all of the earth's oceans, which include an area of about 366 million square kilometers, then there are more than a million submarine volcanoes, and nearly 75,000 of them rise to over 1 kilometer above the ocean floor. It is not known how many are active, but they probably number in the thousands.

Since the end of World War II, advances in the science of remote sensing, advances in drilling techniques, and the development of small submersible vessels able to withstand the high pressures at depths of 6.5 kilometers beneath the sea surface allow exploration of the seafloor by direct observation, by re-mote mapping, and by direct sampling for chemical analyses. The seafloor has gigantic mountains, hills, basins, furrows, volcanoes, and life-forms never be-fore imagined. In the deep rift zones, red-hot lava emerges from cracks and flows across the ocean floor beneath water that is near freezing. The vast ma-jority of volcanoes built along these extensional midocean ridge systems re-main submerged. It is estimated from the known motions of the earth's tec-tonic plates that a total of about 4 cubic kilometers of lava are erupted each year along the 60,000-kilometer length of these systems. None of these erup-tions have been visible above water except where the ridge systems cross onto land in Iceland and in northern Ethiopia and Djibouti.

Not all that comes from midocean fractures is lava. In many areas hot water flows from springs, the first of which were discovered in 1977 at a depth of

Fig. 4-10. Hydrothermal hot spring from black smoker on the deep seafloor gives sustenance to giant clams and worms. (Photo: Courtesy of Rachel Hayman)

2,500 meters along the Galapagos Rift off the coast of South America. Since
then, many hydrothermal springs, *black smokers*, have been discovered.
Black, turbulent clouds with suspended metal-sulfide minerals spout from the
seafloor. Their variable, high temperatures ($\sim 350°C$) produce unique environmental oases capable of supporting a myriad of life-forms—giant clams, mus-
sels, giant tube worms, and a variety of grazers and scavengers (fig. 4-10).
Life-forms also include species of barnacles and limpets thought to have been
extinct for millions of years (since the Mesozoic and Paleozoic Ages). Precipitation of dust-size sulfide minerals from the black smokers builds up mineralized chimneys as high as 13 meters. There are some scientists who claim that
the black smokers were centers around which life on earth may have originated
more than three billion years ago (fig. 4-11).

The mineralized hot-spring mounds have changed our ideas about the origin of some of earth's richest zinc and copper deposits. The release of heat in
the submarine environment may have started about four billion years ago
and initiated the accumulation of sulfides. This is surmised because sulfide

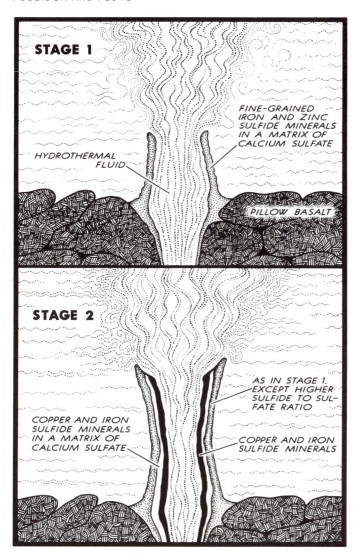

Fig. 4-11. Evolution of a black smoker chimney stack. Plumes of black "smoke" may attain temperatures as high as 350°C. Plumes are actually hot water charged with fine sulphide particles, which build up mineralized chimneys as high as 13 meters. Stage 1. Hot fluids emerge from fractured seafloor to begin accretion of calcium sulfate walls. Calcium sulphate crystallizes from seawater heated around the margins of the jet of hot fluid. The calcium sulfate in the exterior skin of the chimney is partly replaced by fine grains of iron sulfide and zinc sulfide. Stage 2. The formation of calcium sulfate walls protects the hot fluids moving upward through the chimney from mixing with the sea-water; therefore, copper and iron sulfides can crystallize on the inner walls of the chimney. Diameters of chimneys range from about 1 centimeter to 30 centimeters and are usually less than 10 meters high. (Adapted from Haymon and Macdonald, 1985)

deposits in some of the oldest formations on earth, 3.7 billion years old in western Greenland, are similar to modern submarine hot-spring deposits. The ancient formations are found in *ophiolites*, which formed at ancient spreading centers. Many ophiolites have been thrust onto the margins of continents. Large, sulfide-ore bodies in the ophiolites are mined commercially for copper, zinc, and other valuable metals, including gold and silver. Before the discovery of black smokers and the advent of plate tectonics studies, the origin of these ores was an unsolved mystery.

Black Pillows on the Seafloor

An unusual effect that water has on lavas that flow underwater—either in lakes or in oceans—is the formation of "pillows," which are solidified lava buds shaped like tubes, balloons, and, with judicious use of the imagination, pillows (fig. 4-12). As lava moves underwater, its front cools rapidly and slows down, but the flexible skin that develops over the red-hot interior is not strong enough to stop the flow entirely, so its forward movement is accomplished by "budding," which occurs when lava impinges upon weak spots in the plastic skin and causes it to inflate like a balloon. The advancing fronts of flowing lava are not smooth, but instead look like hordes of intertwining, snakelike cylinders with bulbous fronts that mold themselves onto the preexisting surface. If an advancing pillow flow builds a steep front, the pillows inflate and then break off and tumble down the slope. As they tumble, they break into smaller pieces that cool into accumulations of angular bits of basaltic glass called *hyaloclastites*, which are common in Iceland where volcanoes were built under glaciers that have since melted. Heat from the volcanoes melted the ice, and the lava cooled within the pools of water into which it erupted.

Thousands of volcanoes, many with flat tops, dot the ocean bottoms. In tropical regions, some volcanic islands are fringed with coral reefs, some are coral atolls with only a small volcanic tip in the center, and some are coral atolls with no visible central volcano. Atolls are tiny circular islands made of coral reefs enclosing lagoons. In 1831, Darwin theorized that coral atolls are formed as coral grows upward around a slowly sinking volcanic island. In 1947, a little more than one hundred years later, this theory was confirmed at Bikini Atoll when a team of scientists studying the aftermath of the Bikini atom-bomb tests drilled a series of holes into the Bikini reefs. The atoll sits on a 600-meter-thick layer of coral, which rests on the volcanic rock of a submarine volcano. Coral grows only in shallow water, so for such a thick reef to grow, the volcano had to sink as coral continuously built on the upper surface near sea level.

There are a few hundred volcanic islands that have been built above the seafloor. A spectacular example is the island of Hawaii, the Big Island and the youngest island of the Hawaiian-Emperor chain. This chain of 107 volcanoes

Fig. 4-12. A pile of pillow basalts originally formed on the ocean floor but now uplifted and eroded to show a cross section of pillowlike forms. (Photo: Richard V. Fisher)

is 5,600 kilometers long. The volcanoes range in age from 80-million-year-old volcanoes at the northwestern end, near Siberia, to the present-day active volcanoes at Kilauea and Mauna Loa at the southeastern end. Mauna Loa volcano stands 9,300 meters above the floor of the Pacific Ocean. Most volcanoes of the Hawaiian-Emperor chain are volcano remnants that were built 5,000 to 10,000 meters above the Pacific Ocean floor as shoals and seamounts but are now eroded to below water level. Most of the younger end of the chain—the Hawaiian Islands—is severely eroded and flanked in places by spectacular seacliffs. And the Hawaiian-Emperor chain is continuing to lengthen. A new volcano is being built on the sea floor 1,000 meters below the sea surface, 43.6 kilometers southeast of the island of Hawaii. The volcano is named Loihi Seamount and has an edifice 2,700 meters above the floor of the Pacific. If its growth continues at the present rate, it will emerge at sea level in about sixty thousand years. Manned submersibles have been used to observe Loihi since 1987, providing an unparalleled opportunity to watch a seamount grow, but no active lavas have been seen yet.

THE eruptions of Mount Pelée, Mount St. Helens, and Mount Pinatubo, described in part 1, presented some of the destruction generated by volcanoes and how it affects people. A closer look at some of the specific hazards is presented in part 2. The more we know about them, the more success we'll have in mitigating them. Part 2 begins with a look at the fiercely hot pyroclastic flows, which no living thing can withstand.

References

Commission on Volcanology. *Annual Report of the Commission on Volcanology.* Quezon City: National Research Council of the Philippines, 1965–1966.

Cone, Joseph. "Life's undersea beginnings." *Earth* 3 (1994): 34–41.

Darwin, Charles. *Geological Observations.* 3d ed. London: Smith, Elder and Co., 1891.

Dietz, R. S. "The explosive birth of Myojin Island." *National Geographic*, January 1954: 117–28.

Glasstone, S., ed. *The Effects of Atomic Weapons.* Washington, D.C.: U. S. Government Printing Office, 1950.

Haymon, R. M., and K. C. Macdonald. "The geology of deep-sea hot springs." *American Scientist* 73 (1985): 441–49.

Kokelaar, P. "Magma-water interactions in subaqueous and emergent basaltic volcanism." *Bulletin of Volcanology* 48 (1986): 275–89.

Lonsdale, P. "Structural geomorphology of a fast-spreading rise crest: The East Pacific Rise near 3° 25′ S." *Marine Geophysics Research* 3 (1977): 251–93.

Moore, J. G. "Base surge in recent volcanic eruptions." *Bulletin of Volcanology* 30 (1967): 7–363.

Moore, J. G., K. Nakamura, and A. Alcaraz. "The 1965 eruption of Taal Volcano." *Science* 151 (1966): 955–60.

Parks, N. "Exploring Loihi: The next Hawaiian island." *Earth* 3 (1994): 57–63.

Stearns, H. T. *Geology of the State of Hawaii.* Palo Alto, Calif.: Pacific Books, 1966.

Weisgall, J. M. *Operation Crossroads: The Atomic Tests at Bikini Atoll.* Annapolis, Md.: Naval Institute Press, 1994.

PART II

The Hazards of Volcanoes

CHAPTER 5

Volcanic Hurricanes

Frontispiece. Volcanic hurricanes, like those at Mount Pelée in 1902 and at Mount St. Helens in 1980, are one of the most devastating of all volcanic phenomena. They travel faster than ordinary atmospheric hurricanes, can be as hot as the inside of a kiln, and carry tons of ash and rock at high velocities. The force of their winds can blow down forests and destroy buildings. Searingly hot ash particles clog the lungs and throats of victims.

About 35,000 years ago, the Italian coast around Naples was transformed forever within the period of a few hours, and whatever man, beast, or plant that had existed on the land or in the nearby sea on that day was destroyed in a mighty catastrophe—a volcanic hurricane carrying billions of kilograms of rocks and ash hurtled across the Neapolitan landscape. The event was more devastating than any atmospheric hurricane now known to mankind and was propagated by an enormous volcanic eruption centered near present-day Pozzuoli. A volcanic hurricane, which is in no way related to the more familiar atmospheric hurricane, emanates directly from a volcano and, in geological parlance, is a *pyroclastic flow* (*pyro* = "fire"; *clastic* = "broken")—a searing hot cloud that can travel at 160 kilometers per hour or more and carry abundant volcanic ash and particles the size of cobbles and boulders.

The pyroclastic flow produced by the Italian eruption traveled far into the foothills and across the high Apennine mountain ridges. As the ash filled shallow embayments and valleys, new flatlands were formed, including the Volturno Plain and inland valleys beyond nearby mountain ranges. When it was finished, the deposit from the pyroclastic flow (*ignimbrite*) covered an area of 32,000 square kilometers, although at least half that area was in the Tyrrhenian Sea. It is estimated that the original volume of material erupted was 500 cubic kilometers, and the ground was buried to depths of 20 to 100 meters. The new land surfaces along the coast and in the valleys must have been dotted with numerous steam vents as the hot ignimbrite cooled. All of the watercourses and rivers throughout the region were choked with ash debris, and the Apennine mountains were dusted gray with ash over a large area. The Campanian eruption was so large that it affected the land and the sea for hundreds of kilometers downwind from the volcano. Ash spread over the entire Mediterranean Sea and its bordering lands southeast of the Italian peninsula. A layer of ash can still be identified as far east as Crete in sea bottom cores brought up by oceanographic ships.

The Campanian eruption started with magma that rose into a fracture on the north side of the Bay of Naples, believed by some to be in the present-day area of Pozzuoli Bay. With the pressure released, the inflated magma formed a gigantic Plinian eruption column that perhaps extended upward to 40 or 50 kilometers, as shown by the long distance to which the ash traveled eastward when caught by stratospheric winds. The pyroclastic flow formed simultaneously at the base of the eruption column.

As large as it was, the Campanian pyroclastic flow was not an unusual event in the earth's history, nor was it the largest. The earth's surface has been ravaged by thousands, if not millions, of large and small pyroclastic flows. Witnessed volcanic hurricanes have been small when compared with the amount of Campanian Ignimbrite erupted from Pozzuoli Bay, Italy. And as deduced from the great volume of other ignimbrites, even larger pyroclastic flows have moved across the landscape throughout earth's history. The volumes of some indicate the occurrence of eruptions three times larger than the Campanian pyroclastic flow and two to three thousand times larger than the blast at Mount St. Helens. Eruptions of such magnitude would have the potential to destroy all living things over tens of thousands of square kilometers, to deeply bury the landscape, and to send billions of kilograms of volcanic ash into the atmospheric wind corridors of the earth.

Twenty-three thousand years later, the Neapolitan Yellow Tuff was erupted. It occurred at or near the same eruption center as the Campanian Ignimbrite, but it was only one-fourth its size. This later eruption completed the infilling of the Volturno Plain with the products of pyroclastic flows and ash fallout. The yellow cliffs that border the northern suburbs of Naples are the eroded edges of the yellow tuff that covers the Campanian Ignimbrite lying atop the Volturno Plain. The area around Pozzuoli is still volcanically active, the last eruption being in 1538. Studies are currently being made by the Italian government to prepare volcanic hazard maps for the area, and evacuation plans are being made in case of another eruption.

As destructive as their emplacement was at the time, both the Campanian Ignimbrite and the Neapolitan Yellow Tuff have been highly beneficial in the long term. Over the thousands of years since their deposition, weathering has changed their surfaces to a highly fertile soil that has produced abundant crops to feed millions of people. The ignimbrite deposits have also spawned the large building industry in and around Naples. Throughout the last two to three thousand years, blocks of tuff have been quarried by Etruscans, Greeks, Romans, and Italians and used to construct thousands of buildings and walls throughout the region. This industry, still active today, has influenced architectural styles in the Naples region since before Etruscan time (see chap. 11).

Pyroclastic flows were described in some detail at the 1902 eruptions of Mount Pelée, Martinique, by the French volcanologist Alfred Lacroix in accounts that became milestones of volcanological literature. He wrote about the then astonishing fact that instead of going up, the pyroclastic eruption clouds moved along the ground as hot, dense hurricanes (see chap. 1). He named these flowing clouds *nuées ardentes* ("glowing clouds").

Pyroclastic flows can be nearly silent, depending upon the number of collisions of large rocks within them. They can move faster than any observed wind-driven atmospheric hurricane and can reach temperatures nearly as hot as the inside of a kiln. They boil across the ground as a rapidly advancing cloud

Fig. 5-1. Volcanic activity at Mount Unzen, Japan, started in 1991 with the nonexplosive protrusion of a dome on one side of its summit. As the dome grew, the part perched precariously on the edge of the steep slope collapsed. Much to the surprise of volcanologists, instead of creating a rock fall or avalanche, the blocks disintegrated upon collapse to create an ash-laden pyroclastic flow that raced down canyons on the side of the mountain. This photo shows one of the pyroclastic flows as it reached the foot of the slope and continued to spread down the valley. (Photo: Courtesy of Kenji Ohkawa)

with billowing dark lobes that rapidly swell forward to momentarily overwhelm slower-moving lobes, and they send gigantic turbulent updrafts into the sky. They have been seen to move across water, and on land, some have traveled more than 100 kilometers. They are one of the most destructive of volcanic phenomena. More than thirty thousand people have died from direct engulfment by pyroclastic flows since the beginning of this century, most of them at the 1902 Mount Pelée eruption.

Volcanologists first became aware of pyroclastic flows in 1873 at Santorini, Greece, and witnessed them in 1902 at Mount Pelée Volcano, Martinique, and at Soufrière Volcano, St. Vincent. Since then they have been observed and photographed at Mayon Volcano, Philippines, in 1968, 1983, and 1993; at Mount St. Helens, Washington, in 1980; at El Chichón, Mexico, in 1983; at Kelut, Java, in 1983; at Mount Pinatubo, Philippines, in 1991–1993; and at Unzen Volcano, Japan, in 1991–1994 (fig. 5–1).

What Happens inside a
Pyroclastic Flow?

There are two main kinds of pyroclastic flows—one is a fairly dense mixture of ash, stone, and gas and the other is a more dilute gas-rich mixture. Dense pyroclastic flows are confined within and move down valleys. Dilute pyroclastic flows are lighter because they carry fewer fragments, but they can sweep over ridges and into valleys alike, blowing down forests with hurricane-like force, and can move across the surface of water bodies. *Pyroclastic surge* is a synonym for a dilute pyroclastic flow.

The gaseous mixture can carry millions of kilograms of debris across the land and flow 100 kilometers or more. For pyroclastic flows to move long distances on low slopes, the gas needs to be well mixed with the solid fragments and must remain without totally separating from them. The authors maintain that separation of gas from the mixture is slow because small, silt-size particles clog the passageways between larger fragments, thereby hindering settling of the fragments to the ground. Small fragments interfere with the gas-escape routes, therefore gas remains in the mixture, and so long as there is gas within the mixture, the mass will continue to move downhill.

The inside of a moving pyroclastic flow or surge is too dense to be photographed or seen by the human eye. Documentation of the internal movement of particles in the mass is thus circumstantial. Probable movement paths and the manner by which particles fall to the ground and quit moving are deduced from studies on the physical nature of their deposits, from observations of small pyroclastic flows, from computer simulations, and from small-scale laboratory experiments. Theoretical speculations on what should be there are also used, but most important are the characteristics of the deposits left behind.

Natural geological systems such as rivers, glaciers, and air are liquids or gases that can flow without entrained particles. On the other hand, pyroclastic flows are fluids composed entirely of particles mixed with gas. Unlike water or other pure fluids, gravity acts upon the solid particles within the pyroclastic flow and causes the entire mixture to flow. A pyroclastic flow cannot flow without the particles. Without the particles, the hot gases would merely rise into the atmosphere and disperse. Fluids that are a mixture of solids and liquid or gas are called *sediment gravity flows*, the word *sediment* referring to the mass of solid particles.

Imagine that the gases and particles from a volcanic explosion explode sideways from a volcano at high speed and move across the countryside. Gravity pulls at each particle, so it is natural that the larger, heavier ones tend to migrate downward through the roiling mass despite random collisions with other particles that may knock them upward even as they move forward with the mass. The heavier particles filter downward, therefore the upper part of the turbulent current becomes increasingly depleted in particles and therefore more dilute. The dilute flow maintains its turbulence as particles are suspended, but the lower part of the current becomes increasingly more dense and begins to deposit on the ground. Particles continue to hit the upward aggrading surface so the deposit builds from the bottom up and becomes a nonmoving layer. But before the dense lower part of the flow completely stops, the less dense, upper part remains turbulent and separates from the basal part. Therefore, an initially homogeneous flow can develop into two different kinds of flow—the dense one confined to valleys, the less dense (dilute) one capable of surmounting mountain ridges (fig. 5-2).

The deposits left by the passage of dense and dilute pyroclastic flows are different. Those from dense flows are thick, do not have internal layering, and have particles that are larger than those from dilute flows. Deposits from dilute flows are thinner, finer-grained, and layered. These details are important because the different kinds of deposits around old volcanoes can be mapped, telling of the volcano's past—a tell-tale distinction that can save lives, for it provides information on where, how far, and how destructive a pyroclastic flow may have been, and so may foretell what can happen in the future.

What Goes Up Must Come Down

Pyroclastic flows originate in different ways. Some form by the collapse of Plinian eruption columns, like the development of base surges (see chap. 4), others start by the collapse of lava domes, and still others simply boil out of a volcanic crater like a soup pot boiling over.

Plinian eruption columns are composed of hot gases and volcanic particles (see chap. 3), and as the mass rises, it incorporates as much as four times its own weight of air. The heated air reduces the density of the eruption column, which can then rise to great heights. The lower part of the eruption column is more dense than the upper part and commonly collapses downward to form turbulent clouds of hot volcanic ash and pumice that flow away from the base of the column, forming pyroclastic flows. The flows can move faster than hurricane-force winds. The idea concerning the origin of pyroclastic flows from the collapse of Plinian eruption columns (fig. 5-3) stems from watching the collapse of the explosion column that resulted from the detonation of the underwater atomic bomb at Bikini Atoll and from observations of the eruption of Taal Volcano in 1965 (see chap. 4).

Mount Unzen and Merapi

Pyroclastic flows also form from the collapse of lava domes (see chap. 2), as happened in 1991 at Mount Unzen, Japan, and in 1994 at Merapi, in central Java, Indonesia, one of the world's most active volcanoes.

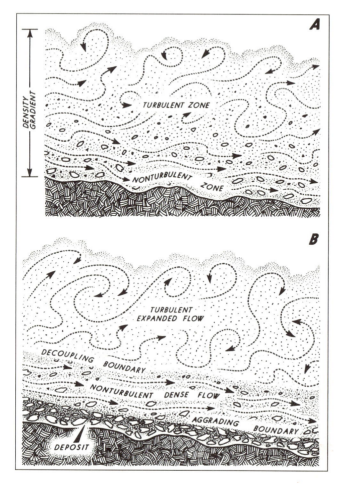

Fig. 5-2. Pyroclastic flows are heavier-than-air mixtures of gases and particles that move at high speeds down the slopes of volcanoes as gravity pulls at each particle. A. Heavier particles migrate downward through the mass and displace lighter particles that move upward with the gases. The heavier mass therefore collects at the base whereas the upper part of the current becomes lighter and more dilute. B. Particles filter down from the dense flow to form a deposit that builds up at the aggrading boundary. Because of the different densities, the lower and upper parts of a pyroclastic flow can separate—the heavier flow confined to valleys, the other capable of surmounting mountain ridges.

Fig. 5-3. Mayon Volcano, Philippines, in eruption on April 28, 1968. The Plinian eruption column extends many kilometers into the atmosphere, while heavier portions at the base of the column collapse to form a pyroclastic flow moving down the flanks of the volcano. (Photo: Courtesy of Arturo Alcaraz)

On June 3, 1991, at Mount Unzen, the American volcanologist Harry Glicken (who had a close call at Mount St. Helens; see chap. 1), the French volcanologists Maurice and Katia Krafft, and forty Japanese journalists were killed by a pyroclastic flow. The killer flow originated at the volcano summit when part of the active dome collapsed, and pressurized, solid fragments broke apart. It was later observed that when such fragments broke, they would explosively produce abundant ash emanating from the fractured surfaces of the very hot pressurized rock. Thousands of such small explosions probably occurred at the newly fractured surfaces as the pressure was released. As the fragments continued to rupture, the combined volume of fine-grained material grew as the mass swept down the valley in the form of a hurricane of hot gases and rock. Glicken and the Kraffts had expected the pyroclastic flow to be smaller, to follow a turn in the valley, and to move past them, following the same path as previous flows. The dense lower part of the flow did follow the valley, but unexpectedly, the lighter-weight upper part of the current separated from the lower part and moved across and up the slope to engulf them as they stood on the opposite side of the valley about 10 meters above the valley floor and about

Fig. 5-4. Volcanic activity at Mount Unzen and pathways followed by pyroclastic flows originating by dome collapse at the summit, beginning May 20, 1991 and stopping in February 1995. On June 3, 1991, the only pathway followed by the flows was the valley (Mizunashi River) extending toward the foreground. (Courtesy of the Nishinippon and Kyushu University Press)

50 to 100 meters from the edge of the former track of a pyroclastic flow that had previously come down the valley (figs. 5-4 and 5-5). The pyroclastic surge that overwhelmed the volcanologists and newspaper reporters was quite dilute and left behind only a few centimeters of ash. Nevertheless, it was powerful enough to push or carry a taxi weighing more than a ton, and hot enough to burn it (fig. 5-6).

Although the pyroclastic flows and surges at Mt. Unzen have been small, they have been remarkably numerous. As of May 1995, between five thousand

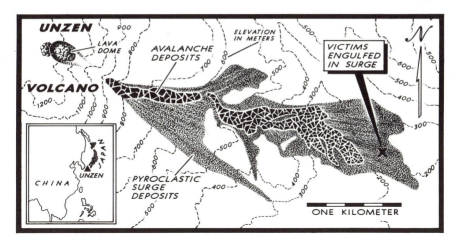

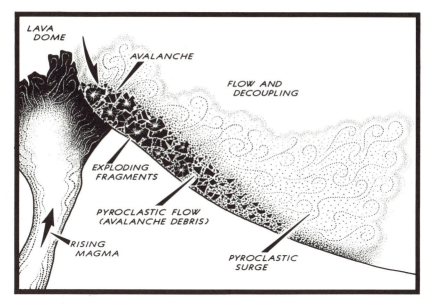

Fig. 5-5. *Top*: A pyroclastic flow from Unzen Volcano on June 3, 1991, killed Harry Glicken, Maurice Krafft, Katia Krafft, and the forty Japanese journalists, who died after being overwhelmed by the hot ash. The basal part of the flow continued down the valley, but the upper part of the flow moved upslope across the valley to overwhelm the victims located in a strategic spot to photograph pyroclastic flows coming down the valley. *Bottom*: The transformation of a dome collapse to a pyroclastic surge.

Fig. 5-6. *Top*: A taxi that was pushed and burned by the pyroclastic surge, June 3, 1991. *Bottom*: The same taxi was pushed for another 60 meters by a dilute turbulent pyroclastic surge on September 15, 1991. (Photos: Courtesy of Setsuya Nakada, Kyushu University)

and ten thousand pyroclastic flows have coursed down the slope of Mount Unzen—according to the Japan Meteorological Agency, there have been more than eight thousand. The Shimabara Volcano Observatory at Kyushu University has counted over fifty-five hundred. But the flows have not been counted visually. Rather, seismometers calibrated to detect pyroclastic flows from their characteristic tremors have been used to determine their number. Differences in numbers reported by the agency and the observatory depend upon the locations of their seismometers.

Dr. Mike Lyvers is a volcano enthusiast who was at Mount Unzen with his wife, Leslie, while on vacation from a postdoctoral position at Kwansei Gakuin University, Japan. He describes what they saw:

> When we arrived [at Mount Unzen] on June 1, the weather was cloudy and rainy, obscuring the mountain. . . . We set up camp on the summit of a high ridge adjoining Mount Unzen, waiting for better weather. A strong smell of sulfur filled the air, and we could often hear the crashing and tumbling of rocks from the new lava dome breaking the eerie silence, but we couldn't see the associated small pyroclastic flows due to thick fog.
>
> On June 2 the weather was unchanged. A newspaper photographer named Yuji arrived at our ridgetop campsite. Yuji spoke English well, and he suggested that we follow him into the evacuated zone at the base of the canyon that night, where all the photographers and film crews were gathered. . . . Yuji said the fiery flows could be seen beneath the layer of cloud. We followed Yuji in his car through the checkpoint, where he told the police we were with him. We then drove through a nice but deserted suburb of the city of Shimabara . . . [and] . . . parked at a bend in the road where dozens of cameras were pointing at the mountain and a large group of reporters were standing around. We got out and looked at the mountain. At first, all I could see was blackness, but periodically the summit area would suddenly light up and what looked like orange-red exploding sparks would tumble down behind an intervening ridge into the canyon, which would then emanate a soft reddish glow. The noise was terrible, a cacophony of avalanching and exploding rocks. Between eruptions it was strangely silent until the terrified dogs started barking again—the evacuees had all left their pets behind. Since we were just below the mouth of the canyon, it was obvious that a large flow could emerge from the canyon mouth and engulf us. I had a terrible feeling of dread and decided to drive out of the area after thanking Yuji. The very next day, as it turned out, the people gathered at that spot [including Glicken and the Kraffts] were killed. (M. Lyvers, e-mail to RVF, April 12, 1996)

Pyroclastic flows and surges occurred on November 22, 1994, at Merapi Volcano, central Java, but gained little coverage in local newspapers of the world. We gleaned the following information from the Internet. The pyroclastic flows and surges developed from the collapse of an active dome at the volcano's summit, similar to what happened at Mount Unzen. Merapi Volcano, 2,911 meters high, is 500 kilometers from Jakarta. The volcano had been active since March 1994, and residents had been warned to evacuate immediately if the activity worsened; however, they had not expected the sudden eruption that occurred in the morning when many were working in their fields. The people knew that Merapi was a dangerous volcano because it had previously erupted in 1930, killing 1,370 people; in 1954 killing 64 people; and in 1976, killing 28 people (see chap. 10). By December 7, fifty-eight people had died, and another twenty-two were in serious condition. All of the victims lived in villages along the Boyong River, which flows south from the summit of Merapi. In addition, fifteen houses were totally destroyed, twenty-five were severely damaged, and six thousand people had been evacuated. Local people had been reluctant to leave the area because they regarded the volcano as sacred and likely to give supernatural signs if it were to cause a major disaster. By early 1995, thousands of people were alerted to the possibility that abundant loose ash from the eruption may continue to spawn dangerous mudflows along the Boyong River during future rainy seasons.

Firecloud Rock:
A Detective Story

Investigations of pyroclastic flows helped solve a long-standing scientific mystery. The 1902 eruption of Mount Pelée initiated worldwide studies of pyroclastic flows and theories on how the flows are able to move large volumes of hot pyroclastic material over the landscape at high speeds. The eruption also helped lead to the explanation for a puzzling kind of rock known as welded tuff, a lava-like variety of ignimbrite. It took seventy years, from the 1860s to the 1930s, to solve the puzzle.

Pure research in the sciences owes some of its advances to technological discoveries, and geology is no exception. Microscopy is one of the techniques used to decipher the origin of rocks (the science of petrology). Rocks are mineral aggregates, but magnification is needed for precise identification. A hand lens is used in the field for a preliminary identification of minerals, but in the laboratory, smaller diagnostic details are made clear by both optical and electron microscopes. Doing so requires special preparation of the rock specimens. A thin sliver of rock is cut by a saw that has industrial diamonds crimped into its edges. The sliver is ground smooth on one side, which is then glued to a glass slide, and the rock chip is further ground thin enough that light can shine through it. Special light-polarizing microscopes were developed to examine such thin sections, and during the nineteenth century, a branch of mineralogy characterized and catalogued hundreds of minerals on the basis of their optical properties. With these tools, petrologists can study the composition and textural relations of minerals and interpret how the rocks were formed.

In the 1860s and '70s, geologists armed with newly developed petrographic microscopes discovered and examined curious lava-like rocks that were found in New Zealand, Armenia, and the United States and were originally thought to be from lava flows. Studies of thin sections showed the rocks to be composed of fragments—in particular, glass shards similar to volcanic ash. The rocks had the composition and hardness of rhyolite lava flows, but some covered hundreds to thousands of square kilometers of land yet measured only tens of meters thick. Such a geometry is highly unlikely for a viscous, silica-rich lava. The earliest and most detailed descriptions of these rocks were made by Ferdinand Zirkel, a noted German petrologist, in 1864. He described pyroclastic features in some New Zealand rhyolites but regarded them as lava flows, thereby contradicting his own observations.

In 1892 an Italian geologist, Luigi dell'Erba, discussed the characteristics and origin of a rock called *piperno*, an extensively used building stone in the region around Naples, Italy, that was believed to be from the Campanian Ignimbrite. He correctly concluded that piperno is a pyroclastic rock deposited at a temperature hot enough to fuse the fragments, and so he deserves credit for first recognition of the origins of piperno and therefore of welded ignimbrite. Zirkel stated in an 1893 publication that Dell'Erba's view was not even probable. A characteristic of piperno is that it contains many fragments of black glass that have a flamelike, wavy shape which the Italians called *fiamme* ("flames") and which Dell'Erba assumed had supplied the heat. Fiamme are pumice fragments that are easily flattened into thin "pancake" shapes by pressure and heat.

While early workers elsewhere argued whether the welded tuffs were lava flows or pyroclastic rocks, the Armenians developed their own ideas about the rock's origin. They postulated that welded tuffs formed from lava flows that expanded explosively after emerging from a vent and then formed a glass froth. This froth then rapidly broke into small sticky pyroclastic particles, and under the weight of the deposited layer, the particles were pressed together and welded. They were called *tufolava* to emphasize their hybrid origin.

The writings of Joseph Iddings, a well-known, late-nineteenth-century American volcanologist, illustrate the ambivalence of researchers at the time and the importance of the 1902–1903 eruption of Mount Pelée, Martinique. In 1899, he indicated that he accepted the Armenian idea of "exploding lava flows" to interpret the origin of Yellowstone welded tuffs. But in 1909, writing six years after the Mount Pelée eruption and after many papers had been published on the origin of nuées ardentes, Iddings omitted the expanding-flow idea and proposed that hot glass fragments may fall together and fuse (weld) into a firm rock. In 1923, Clarence Fenner used the example of Mount Pelée to explain the origin of the Valley of Ten Thousand Smokes, a relatively large ignimbrite in Alaska that had been deposited from a hot "sand flow" from the 1912 eruption of Novarupta (fig. 5-7).

The idea of a nuée ardente as the origin of large lava-like pyroclastic sheets came to Patrick Marshall of New Zealand after reading Fenner's interpretation

Fig. 5-7. In 1912, Novarupta Volcano produced a 12-cubic-kilometer pyroclastic flow (then called a sand flow) that poured down a local valley, depositing a hot, flat-topped mass that filled the valley. Cooling of the hot pyroclastic deposits was facilitated by the formation of hot steam vents (*fumaroles*). There were so many fumaroles that the valley was named the Valley of Ten Thousand Smokes. The pyroclastic flow deposit has largely cooled as of 1995. (Photo: Courtesy of the National Geographic Society; photographer: D. B. Church. *National Geographic*, February 1918)

of the Novarupta (Katmai) sand flow. In 1932, Marshall boldly proposed that the rhyolite plateaus of New Zealand are the deposits of "fiery showers." He ascribed to them a "Katmaian" origin and named them ignimbrite ("fiery rain-cloud rocks"). Marshall's insight and the new name (though poorly defined) sparked the flash of awareness in other volcanologists around the world. Most volcanologists accepted welded tuff as ignimbrite that has a hurricane-like origin similar to that of the Peléan nuées ardentes.

Much has been discovered about ignimbrite since Marshall's time. It is now known that a newly emplaced ignimbrite sheet forms into layers of different hardness or colors because it stays hotter in the center longer than at the bottom or the top. Where it is hottest, the glass shards remain soft longer, become deformed, and weld together. The fused parts of ignimbrite resemble lava flows, and it was the welded tuffs that received all of the early attention. The nonwelded zones above and below the welded zone, though part of the same flow, were considered to be layers formed by other eruptions, and partly for this reason, early efforts to understand the origin of welded tuff had failed. But where did all the ignimbrite come from?

Fig. 5-8. View of Crater Lake Caldera from the west side. The caldera occupies the site of a cluster of volcanoes fed by a magma chamber that extruded so much material that it created a subterranean cavity into which the volcano field collapsed. This created an 8-kilometer-diameter crater that collects rain water to form the lake. A late gasp of volcanism built the cinder cone known as Wizard Island on the west side of the crater. (Photo: Grant Heiken)

Calderas and Pyroclastic Flows

It took geologists many years to figure out the obvious—that the eruption of immense volumes of volcanic ash and pumice to form ignimbrite empties an underground void, probably momentarily, equal to that of the erupted mass. Rocks that lie above the void collapse into it, thereby creating large craters at the earth's surface. Such craters are called calderas (see chap. 2). It is axiomatic that the presence of a large-volume ignimbrite deposit indicates that somewhere there is a source caldera—and the larger the ignimbrite sheet, the larger the crater. Because "mega-eruptions" eject enormous amounts of material within days or weeks, they create huge calderas. Because the calderas are so big, most people who visit geothermal features such as those of Yellowstone National Park are unaware that they are within some of the largest volcanic craters on earth. Even geologists need prior knowledge to recognize such immense calderas. But there are also smaller ones, ranging from 1 to 10 kilometers in diameter, where it is possible to see the entire crater (fig. 5-8).

Three major volcanic stages in the making of a caldera have been discovered. First, small volcanic eruptions occur, followed, second, by culminating

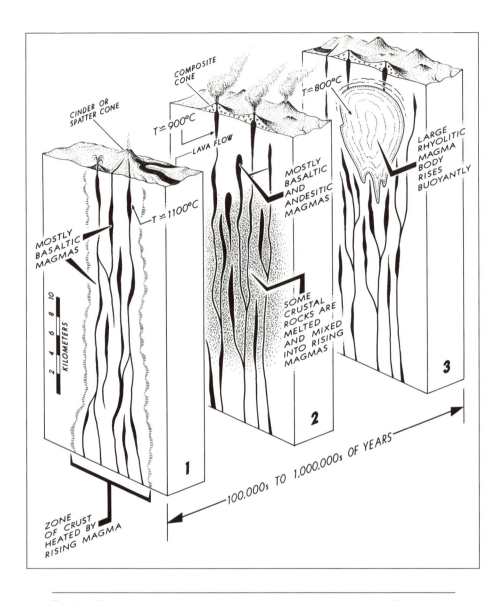

Fig. 5-9. Slices deep into the earth's crust depict the caldera cycle at different times. *Above*: The rise of the magma. At time 1, basaltic magmas have risen to produce small cinder cones. The rising basaltic magma heralds the rise of a large magma source. At time 2, andesitic magmas have risen with the basaltic magmas to build composite volcanoes. At time 3, a large body of rhyolitic magma has risen into the upper crust. *Right*: The eruption. At time 4, the rhyolitic magma erupts, creating a caldera as pyroclastic flows spread out from the source. At time 5, the pyroclastic flow has been deposited, and the floor of the caldera has domed up slightly by upward pressure of the magma body. Not shown are small lava domes that form around the rim of the caldera or within the crater, which commonly ends the cycle. Cycles are measured in millions of years.

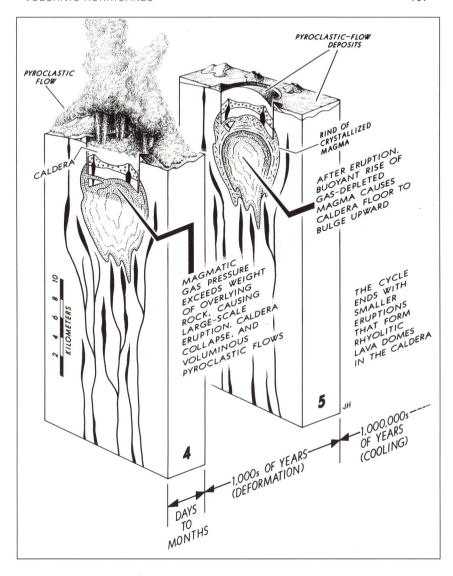

PYROCLASTIC-FLOW
DEPOSITS

PYROCLASTIC
FLOW

RIND OF
CRYSTALLIZED
MAGMA

CALDERA

AFTER ERUPTION,
BUOYANT RISE OF
GAS-DEPLETED
MAGMA CAUSES
CALDERA FLOOR TO
BULGE UPWARD

10
8
6
4
2
KILOMETERS

MAGMATIC
GAS PRESSURE
EXCEEDS WEIGHT
OF OVERLYING
ROCK, CAUSING
LARGE-SCALE
ERUPTION, CALDERA
COLLAPSE, AND
VOLUMINOUS
PYROCLASTIC FLOWS

THE CYCLE
ENDS WITH
SMALLER
ERUPTIONS
THAT FORM
RHYOLITIC
LAVA DOMES
IN THE CALDERA

5
JH

1,000,000s
OF YEARS
(COOLING)

4

1,000s OF YEARS
(DEFORMATION)

DAYS
TO
MONTHS

eruptions of voluminous pyroclastic flows slightly before and at the same time as the caldera collapses. Third, smaller postcollapse volcanic eruptions may continue intermittently for a million years or more (fig. 5-9).

The distribution of volcanic ashfall deposits over many western states (fig. 5-10) are evidence that eruption columns from Yellowstone entered the stratosphere. The volumes of glass-rich volcanic ash that have spewed from the calderas now occupied by Yellowstone National Park include 2,500 cubic kilometers, 2.2 million years ago; 280 cubic kilometers, 1.2 million years ago;

Fig. 5-10. Three eruptions from what is now Yellowstone National Park have been truly enormous. Over the last 2.2 million years these eruptions collectively produced 3,800 cubic kilometers of ash and pumice. Measurable ash fallout from these eruptions covered a substantial part of the conterminous United States and parts of Canada.

and 1,000 cubic kilometers, 600,000 years ago (fig. 5-11). These figures include the ignimbrite as well as the ashfall across the Midwest. Considering that the eruption of Mount St. Helens in 1980 created so much destruction with only 2.5 cubic kilometers of erupted ash and rock, unimaginable environmental damage must have gripped the western United States at the time of the prehistoric Yellowstone eruptions, for they were about one thousand times larger than the Mount St. Helens episode. During the last two million years of gigantic caldera eruptions, starting with Yellowstone, there were three others in North America—two in the Jemez Mountains of New Mexico near

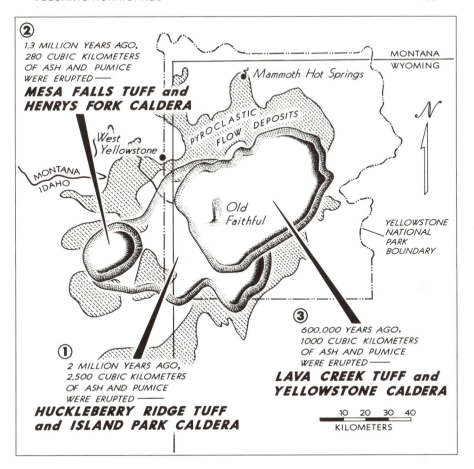

②
1.3 MILLION YEARS AGO, 280 CUBIC KILOMETERS OF ASH AND PUMICE WERE ERUPTED—
MESA FALLS TUFF and HENRYS FORK CALDERA

Mammoth Hot Springs

MONTANA
WYOMING

West Yellowstone

PYROCLASTIC FLOW DEPOSITS

MONTANA
IDAHO

Old Faithful

YELLOWSTONE NATIONAL PARK BOUNDARY

③
600,000 YEARS AGO, 1000 CUBIC KILOMETERS OF ASH AND PUMICE WERE ERUPTED—
LAVA CREEK TUFF and YELLOWSTONE CALDERA

①
2 MILLION YEARS AGO, 2,500 CUBIC KILOMETERS OF ASH AND PUMICE WERE ERUPTED—
HUCKLEBERRY RIDGE TUFF and ISLAND PARK CALDERA

10   20   30   40
KILOMETERS

Fig. 5-11. At 40 to 60 kilometers in diameter, Yellowstone Caldera is one of the largest calderas in the world, yet it forms part of a caldera cluster consisting of two others, Henry's Fork Caldera and Island Park Caldera. These calderas and their ignimbrite sheets were formed 2 million (Henry's Fork), 1.3 million (Island Park), and 600,000 years ago (Yellowstone). Together, the three calderas have ejected 3,800 cubic kilometers (960 cubic miles) of volcanic material.

Santa Fe 1.4 and 1.1 million years ago, and one at Long Valley near Bishop in eastern California, seven hundred thousand years ago. The largest volume of ignimbrite measured so far (in the San Juan Mountains, Colorado) is over 3,000 cubic kilometers. Eruptions with volumes of 500 cubic kilometers occur about every one hundred thousand years. Not only would another Yellowstone event the size of the last one be disastrous due to the outpouring of ignimbrite, but far downwind, the large volume of volcanic ash would cause havoc for years. We know that this is possible because the midcontinent of the United States and Canada has been the ashfall site for three of the six largest volcanic eruptions in North America during the last two million years, all from calderas.

Even though eruptions that form calderas can be gigantic, only a part of the magma body is erupted—no more than about 10 percent of the total. Underlying the volcano (or volcanic field) at depths of two to seven kilometers is the larger residual volume of molten material that takes hundreds of thousands of years to cool from the original temperature of about 1,000°C to the ambient temperature of the surrounding rocks. Yellowstone, with its widespread geothermal features of hot springs, geysers, and fumaroles, is the surface manifestation of a huge regional underground reservoir of cooling magma.

There is an undercurrent of apprehension exhibited by the volcanologists who study the deposits and calderas of truly large eruptions. Judging by problems from the Mount St. Helens ashfall in eastern Washington, an eruption one thousand times greater could be a cataclysm beyond imagination. We can postulate that another gigantic Yellowstone eruption would paralyze transportation, communication, and energy systems throughout North America and could affect the world's weather patterns and inhibit the growing season. It would take many years to clear the roads and to replace electrical lines, generators, and switchyards damaged by coatings of fine ash. Swirling clouds of volcanic dust would disrupt cities and decrease food supplies, and without electricity, existing food supplies could not be preserved. Air filters of engines would clog and their moving parts would be quickly abraded. The potential of airplanes to fly would probably not be harmed if all the ash is on the ground, but they would not be able to land or take off from ash-covered runways. The problems would be compounded by rain, which transforms ash to slickened mud on highways and all other surfaces. Also, wet ash would place a crushing force on all structures. And with such a tremendous quantity of ash, it would be difficult to find a place to put it. The landscape of flat prairielands would be changed forever, dotted by great mounds and hills of ash scraped from streets, roads, airports, etc. During winter seasons, rivers and streams would become torrents of mud. The countryside would look as it does after a heavy snowfall, but nothing would melt in the spring.

According to the present distribution of volcanic ash from the largest Yellowstone eruption, another one of comparable size would deposit ash over much of the headwaters of the Missouri and Mississippi Rivers (see fig. 5-10). Such a voluminous ash fallout would have a significant impact. Continued stripping of ash from the land into the Missouri-Mississippi river system could increase the severity of annual floods for decades and cause many of the barge channels to silt up. River waters containing abundant fine ash could no longer be used to cool water for the many electricity generating plants along the Mississippi because filters could be clogged. All the way to New Orleans and out to the Mississippi delta and Gulf of Mexico, the river would be a milky plume of fine debris that would be visible from space.

The most immediate global effect of such a large eruption would be on the world's food supplies. Cereals make up half the caloric intake of the people on earth. In 1986, according to the United Nations Food and Agricultural Organization, more than 1.8 billion metric tons of cereals were produced by the world's farmers. Half of that was grown in North America, and much of it within the region downwind from Yellowstone. In addition to the disruption and flooding of the midcontinent river system, an eruption similar to Yellowstone's last caldera-forming event could leave an ash layer many centimeters thick on much of the continent's highly productive farmland and thereby directly reduce cereals production.

While this science-fiction-like scenario of a Yellowstone eruption is possible, it is not likely that anybody alive today or their grandchildren's grandchildren will experience such an infrequent volcanic event. The extraordinarily large eruptions have left a positive legacy in providing geothermal energy for heat and electricity and valuable mineral deposits. If the frequency of such large eruptions is about every one hundred thousand years, mankind has experienced only ten of them in the last million years, but there is no recognizable human written or oral record of them. Given earth's longevity, however, there is time for many more to happen.

Volcanism after a Caldera Collapse

The length of the life cycle of a caldera is beyond the comprehension of human beings, whose point of reference is a life cycle of forty to one hundred years. The formation of magma and its rise upward through the earth can take millions of years, but the climactic eruption does not end its life cycle. After a caldera collapses, eruptions may occur off and on, in and around the caldera for millions of years. Small volcanoes may grow in the crater, or domes may grow around the margins as magma leaks through fractures made in the crust during the collapse. Mammoth Mountain, with the town of Mammoth at its foot, is a popular tourist resort that is also a dome complex built on the rim of Long Valley Caldera. The mountain possibly originated from magma rising along the caldera's ring fracture.

The life of a caldera after it collapses is generally not as vigorous as before, but the magma is still active and may dome up caldera floors as it rises. Such resurgent doming occurred within Long Valley Caldera near Mammoth to form a domelike group of small hills within the caldera.

World War II and the Battle for a Caldera

A notable resurgent mountain is the island of Iwo Jima, in the Pacific Ocean, which is the deformed floor of a submarine caldera. The stirring photograph of five marines raising the American flag on Mount Suribachi, one of the highest elevations of Iwo Jima, will ever symbolize the valor and bravery of the young men of World War II. The barren island of Iwo Jima became famous because of the terrible struggle to take it from the Japanese in a battle that lasted from February 19 to March 26, 1945. The toll was terrible—26,800 young men died: 20,000 Japanese and 6,800 Americans (fig. 5-12).

Iwo-Jima, also called Sulfur Island, is located about 1,150 kilometers south of Tokyo and is one of the many volcanic islands that extend from the Mariana Islands north to Tokyo Bay. These islands were the stepping stones used by Allied forces as forward bases to conduct the war with the Japanese in 1944 and 1945. Iwo Jima was along the flight path for B-29s traveling from the Marianas to Japan. It had radar, was a base for the Japanese fighter planes, and was therefore a problem for the Army Air Force. It also had two air strips that the United States was desperate to acquire for emergency landings and for a refueling station; the B-29s could not carry enough fuel for the round trip from the Marianas to Japanese cities if headwinds were strong. The battle for Iwo Jima was costly and the volcanic nature of the island played a part in the battle.

In 1985 a team of geologists led by Sohei Kaizuka of the Tokyo Metropolitan University published the results of their combined geologic and oceanographic studies of Iwo Jima and the sea around the island. They demonstrated that the island of Iwo Jima is the emerged part of the uplifted floor (structural resurgent dome) in the center of a ten-kilometer-diameter submarine caldera formed 2,600 years ago. Magma is moving upward and being injected into the crust three to four kilometers below the caldera floor, raising it to form the island. This uplift has continued intermittently for at least the last 700 years, and its rate of rise, based on carbon-14 dates of wave-cut terraces, has been 15–20 centimeters per year, a phenomenal pace by geological standards. This rapid uplift is easily confirmed by measuring the heights of the wave-cut shorelines. In 1779, Captain Cook's surviving crew landed on a shore that is now 40 meters above sea level. U.S. naval landing craft that wrecked on the beach during the 1945 battle are now nearly 10 meters above the present-day beach.

Further inland, the small volcano known as Mount Suribachi consists of interbedded lavas and tuff that were erupted from within the caldera. At the time of the battle, it was (and still is) a barren knob covered with pumice. The hot fumaroles scattered across the island are obvious signs of active volcanism. As an example of the amount of heat emanating from the ground, the drilling of a water well had to be abandoned because the heat at shallow depths took the temper out of the drill stem and drilling could not continue. Moreover, since 1889, when the Japanese began keeping records, there have been sixteen crater-forming steam blasts. The last one occurred in 1957, when a sixty-five-minute eruption blasted a 35-meter-diameter, 15-meter-deep crater alongside an abandoned World War II runway. It was reported that the eruption occurred "without warning," but U.S. Air Force personnel had noticed that the floor and

Fig. 5-12. Landing craft clustered near the shores of Iwo Jima to unload men onto the soft, pumice-rich sands, which impeded their forward progress against the entrenched Japanese. Iwo Jima is a barren and lonely volcanic dome within a submerged caldera in the Pacific Ocean south of Japan. The remote island was needed by U.S. forces as an advanced air base by which to wage war against the Japanese in World War II. The invasion by U.S. Marines started on February 19 and continued to March 26, 1945, during which 20,000 Japanese and 6,800 Americans died. (Photo: U.S. National Archive photo 80-G-307182)

cold-water supply of one of the base buildings had become naturally heated in 1955—an unheeded precursory warning. The only casualties of the steam eruption were some sea birds and field mice.

The northeastern part of the island is a 115-meter-high plateau built of tuff, which originally filled the caldera. But now, the originally flat-lying tuff deposits have been bowed upward by magma under the volcano. Wave-cut beach terraces, like bathtub rings, were formed by the erosion of soft volcanic ash to make beaches that were lifted by the rising dome, and these occur at many levels around the island's perimeter. Because resurgence recurs sporadically, there is enough time for shoreline waves to cut a small beach terrace before the next rise of the island occurs. The rate of uplift can be calculated by knowing the age and height of each beach-terrace level.

The battle for Iwo Jima was influenced in large measure by these physical volcanological attributes of the island; yet another example of the interaction of volcanoes and human endeavor—this time, concerning the vagaries and tactics of war. The beach terraces, like a series of large stair steps, were mounted by the U.S. forces during the battle, but only with great difficulty because the steps were mantled by loose volcanic ash, pumice, and sand. Traction was nearly impossible for the tanks, trucks, and foot soldiers coming ashore to engage in battle. Moreover, the rather soft tuff comprising the island's central plateau had been easily excavated into an intricate defensive network of tunnels by the Japanese forces. As a result, the Japanese Forces were extremely difficult to dislodge.

The battle for Iwo Jima was for a speck of land that constantly changes character because of the dynamic processes that can continue within young calderas. At the moment the island is growing; however, should the subterranean magma break through to the surface, it is possible that the rising island could disappear forever in a cataclysmic eruption.

PYROCLASTIC flows and caldera formation can devastate huge land areas. Not as widespread, but just as deadly on the slopes and river courses of volcanoes, are the spectacular falling mountains and widespread mudflows and floods that will be discussed in the next chapter.

References

Cook, E. F., ed. *Tufflavas and Ignimbrites*. New York: American Elsevier, 1966.

Dupras, D. "The grandeur of concrete: Part I." *Oregon Geology* 51 (1989): 37–44.

Fisher, R. V. and H.-U. Schmincke. *Pyroclastic Rocks*. Berlin: Springer-Verlag, 1984.

Francis, P. *Volcanoes: A Planetary Perspective*. Oxford: Oxford University Press, 1993.

Hooker, Marjorie. "The origin of the volcanological concept *Nuée ardente*." *Isis* 56 (1965): 401–7.

Lipman, P. W. "The roots of ash flow calderas in western North America: Windows into the tops of granitic batholiths." *Journal of Geophysical Research* 89 (1984): 8801–41.

Falling Volcanoes and
Floods of Mud

Frontispiece. The telltale mud line on trees along the Toutle River, Washington, shows the fury and extent of the flood of mud that overfilled its banks at peak flow about four hours after the collapse of Mount St. Helens. The man leaning against tree indicates the height of the mud line at a point 65 kilometers downstream from Mount St. Helens Volcano. (Photo: Courtesy of Harry Glicken)

A Mountain Fell

On Sunday morning May 18, 1980, a mountain fell. The day was sunny and cloudless and had dawned on a seemingly quiet Mount St. Helens. Keith and Dorothy Stoffel, both geologists, had chartered a small airplane on a Sunday outing to take photographs of Mount St. Helens; they were over the volcano at 8:32 A.M. when it collapsed. They were closer than anyone, with an unparalleled bird's-eye view, and had circled the mountain three times, taking photographs. On the fourth pass, they saw rock and ice debris landslide into the crater, and within fifteen seconds, the north side of the volcano began to ripple, quiver, and churn and then began to slump and flow toward the valley. As they photographed the slide, a cloud jetted violently to the north from a scar left barren by the mass that had just slid away. Within seconds the cloud had expanded to such dimensions that their view was obscured because the blast was directed straight toward them. With swift instincts, the pilot opened full throttle and dove steeply downward to quickly gain speed, but though they were moving at about 200 knots, the blast cloud gained on them. It was obvious to the pilot that the expanding cloud was moving fastest to the north, so he turned south, and they escaped certain death.

On the ground, many volcano watchers, with cameras ready, waited for the volcano to erupt. One viewpoint was at Bear Meadow, 18 kilometers northeast of the volcano. Had someone written a fictional account of a collapsing volcano, they could not have imagined a better locality from which to observe the eruption. One main character was an amateur photographer named Gary Rosenquist. He had his camera set on a tripod pointing high at the volcano to photograph an expected eruption column, but his wife had tripped slightly on the tripod. This accidentally lowered the camera's aim toward the bulge in Mount St. Helens, in exactly the right direction to photograph the volcano's collapse, the avalanche, and the northward moving blast.

Like a miracle, the conditions were optimal for photography, because in May it is usual for the Washington Cascades to be covered in clouds. And to complete the miracle, both Rosenquist and his photographs survived. Miraculous because the blast that followed the avalanche bore down toward Bear Meadow at about 200 kilometers per hour but stopped 1.6 kilometers short of Rosenquist's location. About 1.6 kilometers to the west, another lobe of the blast continued northward past Bear Meadow for another 7.2 kilometers.

Rosenquist had captured the event at timed intervals of twenty seconds to preserve an enormous amount of scientific data that could never have been described in words.

It was an astonishing event. The avalanche on the north side of the volcano plunged 2,000 meters into the North Fork of the Toutle River at 250 kilometers per hour. As it crossed the Toutle River, it encountered a high ridge and divided into three parts. The eastern branch flowed into Spirit Lake, causing a gigantic wave that washed up the slopes as far as 76 meters like the swash in a suddenly loaded bathtub. Directly northward, its momentum carried the avalanche over a 300-meter ridge, later named Johnston Ridge, and down the other side into the small east-west valley of South Coldwater Creek, where it continued to flow down the creek. The main western branch slammed into Johnston Ridge, was diverted down the North Fork of the Toutle River, and flowed down the river for 23 kilometers. Three cubic kilometers of the volcano had been transformed into a debris avalanche that covered 60 square kilometers.

After Harry Glicken's supervisor, David Johnston, was killed by the blast directed from the north side of Mount St. Helens, Glicken dedicated himself to an investigation of volcanic sector collapse. He found the deposit of the debris avalanche to be a complex of rock fragments consisting of one-fourth of the volcano—a jumbled mosaic composed of very large blocks in close contact with one another, and smaller and smaller pieces, down to microscopic fragments.

Glicken divided the volcanic debris-avalanche complex into blocks and "matrix" (the fine-grained broken rock between large blocks). The blocks are relatively intact pieces of the old volcano transported several kilometers, but many of the original ones were whipped across the ground surface at such high velocity, that they became "smears" made of thousands of small pieces of the original block. The matrix surrounding the blocks is a loose mixture of all the rock types from the old mountain, some from the freshly exploded magma and some picked up from the ground. The hillocks scattered around on the surfaces of debris avalanches are usually coherent blocks that "float" in the matrix of smaller pieces of the broken mountain. But at Mount St. Helens, Harry Glicken found more than one kind of hummock. One type is constructed of blocks that touch one another without matrix material between them. Some of the blocks are completely shattered rock of a single type, and others are composed of many types. The least broken blocks can be reconstructed into a whole, like a jigsaw puzzle, because the broken pieces are not far removed from one another. Others are so highly shattered that samples of the once hard lava rock can be removed with a spoon and sieved like sand.

The highly distinctive hilly, or hummocky, surface found at the Mount St. Helens debris avalanche had been previously described at Galunggung Volcano, Indonesia, and at Bandai San Volcano, Japan, but few volcanolo-

Fig. 6-1. A view of the barren and desolate land that replaced a verdant forest as deposits filled a valley on the cataclysmic day of May 18, 1980, when the north side of Mount St. Helens collapsed. The mountain stands in the background with its horseshoe-shaped crater open to the north. The pieces of the volcano that slid and avalanched are the hummocks at the foot of the volcano and in the foreground. (Photo: Richard V. Fisher)

gists were aware that the hills were caused by avalanches. Some interpreted the hummocky ground to be the result of large volcanic mudflows. Glicken's work detailed the mechanisms and little-known internal physical characteristics of volcanic avalanche deposits for the first time. The collapse of Mount St. Helens dramatically demonstrated that volcanoes cannot grow to an unlimited height within the earth's gravitational field. They will eventually collapse under their own weight, causing avalanches.

Some of the blocks at Mount St. Helens are more than 100 meters across and stand in hills or hummocks more than 70 meters above the level of the flow's surface (fig. 6-1). In other much older volcanic avalanches, such as at Mount Shasta, hummocks are over 100 meters high, and at Nirasaki, Japan, blocks within an ancient volcanic debris avalanche are as long as 1,500 meters.

During the July 15, 1888, eruption of Bandai San Volcano, Japan, its north side collapsed to create a gigantic avalanche. Sekiya and Kikuchi, two

Japanese volcanologists, reached the mountain within days of the eruption. In 1890 they reported what they had witnessed:

> At about half-past seven, there occurred a tolerably severe earthquake, which lasted more than 20 seconds. This was followed soon after by a most violent shaking of the ground. At 7:45, while the ground was still heaving, the eruption of Kobandai-san took place. . . .
>
> The last explosion . . . is said to have projected its discharge almost horizontally, towards the valley on the north. . . . The main eruptions lasted for a minute or more, and were accompanied by thundering sounds which, though rapidly lessening in intensity, continued for nearly two hours. Meanwhile the dust and steam rapidly ascended. . . . At the immediate foot of the mountain there was a rain of hot scalding ashes, accompanied by pitch darkness. . . . While darkness as aforesaid still shrouded the region, a mighty avalanche of earth and rock rushed at terrific speed down the mountain slopes, buried the Nagase valley with its villages and people, and devastated an area of more than 70 square kilometers, or 27 square miles. (Sekiya and Kikuchi, "The eruption of Bandai-san," p. 103)

The event was re-created as a fictional account, but with historical accuracy, by the distinguished Japanese writer, Inoue Yasushi:

> At exactly 7:40 the earth gave a great heave and shudder. This was different from the tremors we had felt before, much more violent, and I was knocked to the ground. I could not tell if it came from the mountain or the ground beneath me, but I heard the most terrifying sound issuing from the bowels of the earth. I saw the young woman lose her balance, stagger, and fall to her knees. I scrambled to my feet only to be thrown down again by a second violent jolt. This time I used my right arm to brace my body against the bucking earth. I glanced up at the outcrop to see if the children had also been thrown down, but there was no sign of them. All I could see was a swirl of dust slowly rising in the air.
>
> By this time I knew better than to try and leap up again, but after the second quake subsided, I carefully rose to my feet. Beside me I saw that the young man had reached out a hand and was helping the woman up as well.
>
> At the same moment, I saw two or three small heads poke up above the edge of the outcrop. Soon all the heads appeared in a row and I heard one of the children cry out in a loud voice, cadenced almost as though he were singing, "Blow, mountain, blow! Give it all you've got!" Soon several of the others joined in, shouting with all their tiny might, "Blow, mountain, blow! Give it all you've got!"
>
> Their chant—or scream of defiance, whatever it was—was scarcely finished when in thunderous answer a roar came rolling back over the earth. It was a blast so powerful that I was lifted off my feet and hurled to the ground several yards to my right. On and on went the roar while the earth heaved in convulsive spasms. Later when I tried to recall the exact sequence of events, I was never sure just when it was that I happened to catch sight of Mt. Bandai, but I know I saw a huge column of fire and smoke rising straight up into the clear tranquil sky; like one of the pillars of Hell

it rose to twice the height of the mountain itself. The whole mountain had literally exploded and the shape of Kobandai was blotted out forever. It was only much later, of course, that I realized what had occurred.

I cannot say with any certainty how I survived the explosion. The entire north face of Mt. Bandai came avalanching down in a sea of sand, rocks, and boulders. I remember it now as a nightmare vision, as something so terrifying as to be not of this world. The avalanche obliterated the forests that covered the lower slopes of the mountain. The wall of debris swept down with terrible speed and force. . . . The air was so thick with clouds of ash and pebbles I could not tell whether it was day or night. I staggered along the bank of the Ono River and sought refuge on the high ground north of Akimoto. That alone saved my life. If I had fled in any other direction I would simply have been whisked away without a trace.

Within an hour of the time Mt. Bandai exploded, the villages of Hosono, Osawa, and Akimoto were all swept away, and whatever remained was buried. . . .

What remains indelibly burned upon my memory and ringing in my ears is the defiant challenge—"Blow, mountain, blow! Give it all you've got!"—uttered by those brave children, who could do nothing else in the face of the mountain's awesome power.

Today Hosono, Osawa, and Hibara are all buried beneath the large lake that formed when the stones and mud of the eruption blocked the Nagase River. Akimoto lies at the bottom of another such lake. Though I have related this story in some detail, the fact is that I have never gone back to visit the area, and it is unlikely that I ever shall. The region today, they tell me, is noted for its pristine alpine lakes, but who can say what terrible memories would revive if I were to go there again and gaze upon them. No, I shall never revisit the countryside that lies in the shadow of Mt. Bandai. (Yasushi, "Under the Shadow," pp. 252–54)

Several volcanic debris avalanches within historic times were larger than one cubic kilometer in volume (table 6–1). Four have occurred in this century, and since the events at Mount St. Helens, several hundred ancient volcanic debris avalanches have been discovered throughout the world. No known one has yet swept across a highly populated area, but one that entered the sea from Unzen Volcano in 1792 created a tsunami that killed 14,524 people in nearby coastal villages. The area north of Mount Shasta Volcano in northern California, dotted with dozens of hills and mounds formed 350,000 years ago, is a gigantic avalanche deposit. It covers 450 square kilometers and has a volume of 26 cubic kilometers (fig. 6-2). Prior to the avalanche at Mount St. Helens, the Mount Shasta avalanche deposit was thought to be a gigantic mudflow deposit or erosional remnants of older volcanoes; we now know that it is the remnants of a large avalanche that occurred during the collapse of an earlier Mount Shasta.

Most known debris avalanches occurred prior to written history. One of the largest and oldest known is a sixty-million-year-old avalanche deposit in the Absaroka Mountains, Wyoming, known as the Castle Crags Chaos, which was

Table 6-1. Large-Volume Slope Failures of Volcanoes since A.D. 1600

Volcano	Location	Year	Volume km³	Area km²	Fatalities*
St. Helens	Cascades	1980	2.5	57	64 (b,a,l)
Shiveluch	Kamchatka	1964	1.5	98	none known
Bezymianny	Kamchatka	1956	0.8	30	none known
Bandai San	Japan	1888	1.5	34	461 (a,b)
Unzen	Japan	1792	0.34	15	14,524 (t,a)
Papandayan	Indonesia	1772	0.14	18	2,957 (a)
Oshima-Oshima	Japan	1741	0.4		1,475 (t)
Iriga	Philippines	1628?	1.5	65	probable (a)

  * a = avalanche; b = blast surge; l = lahar; t = tsunami.

Fig. 6-2. The hills in the foreground are the remnants of a 350,000-year-old avalanche produced by the collapse of Mount Shasta. The volcano is seen in the distance, 64 kilometers to the south. The avalanche originally covered 450 square kilometers, many times larger than the one from Mount St. Helens. (Photo: Courtesy of Harry Glicken)

the subject of a doctoral dissertation by Kent Sundell in 1985, when he was a colleague of Harry Glicken at the University of California. This ancient deposit covers no less than 1,550 square kilometers—larger in area than the city of Los Angeles—and contains single blocks as long as 1.5 kilometers.

Debris Avalanches and Slides beneath the Sea

Debris avalanches also take place beneath the sea along the edges of islands, a potential risk for islanders, particularly in the Hawaiian Islands. If the water were to disappear from the Pacific Ocean, the Hawaiian Islands would reveal their huge dimensions. They are part of the longest volcanic chain on earth and are the tallest mountains on the planet (see chap. 3). It should then come as no surprise that there are huge debris avalanches commensurate with the size of the volcanoes. Off the northeastern shore of Oahu is a gigantic landslide of jumbled rocks that covers 23,300 square kilometers and extends for 242 kilometers across the ocean floor. In places it is 1,610 meters thick—the largest known landslide on earth. Parts of each of the Hawaiian Islands have collapsed, for they are water-saturated, heavy, and stand thousands of meters above the ocean floor.

The island of Hawaii is surrounded by gigantic landslides off its shores. Because the Big Island is still growing, it can be predicted that immense slides will occur in the future. It is especially important to understand the causes of the landslides because they slough off great tracts of land like a giant bite that can reach back for kilometers into the island, and thousands of people on the Hawaiian Islands live within 10 kilometers of the coastal region. Large submarine landslides are also known to cause tsunamis. Tsunami deposits at 200 meters above sea level on several of the Hawaiian Islands suggest that gigantic landslides have occurred in the past. Fortunately, such monstrous landslides and resulting tsunamis do not occur very often, and no one in Hawaii has experienced them in historic time.

At Mount St. Helens, some of the volcanic floods and debris flows originated from the avalanche deposits, and so we will discuss them here, even though they usually form in other ways. The one similarity between debris flows and avalanches is that they are made of broken volcanic fragments. Debris flows are lingering hazards around volcanoes, and following some eruptions, they can continue many years.

Volcanic Debris Flows (Lahars)

On the night of November 13, 1985, while most citizens of Armero, Colombia, lay asleep, a killer debris flow (lahar) suddenly emerged from the narrow confines of Lagunillas Canyon. It had come from Nevado del Ruiz Volcano after

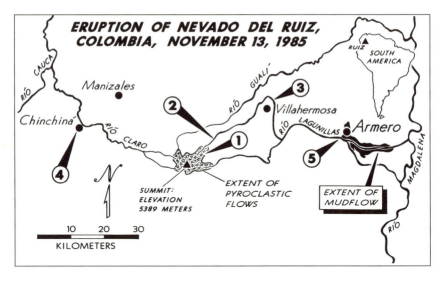

Fig. 6-3. At the 5,389-meter summit of Nevado del Ruiz on November 13, 1985, an eruption of four hot pyroclastic flows melted snow and ice. The water plunged downward into canyons accumulating mud and becoming lahars in all of the rivers leading away from Nevado del Ruiz. The most destructive lahar traveled down Rio Lagunillas. Within two hours it raced 70 kilometers, dropping 4,389 meters in elevation and emerging from its canyon to demolish the town of Armero. Site 1. 9:08 P.M.—eruption of four pyroclastic flows, melting of summit snow pack. Site 2. 10 P.M.—10 kilometers downstream, peak flow of 48,000 cubic meters per second. Site 3. 10 A.M.—mud waves 30-40 meters high. Site 4. 10:30 P.M. Chinchiná struck by first of many lahar waves (flow rate is 13,000 cubic meters per second). Site 5. 11:30 A.M.—the first pulse of lahars hits Armero; lahar emerges from ravine above town with a flow rate of 27,000 cubic meters per second and a 40-meter-high wave.

a small eruption, then spread across the alluvial plain and swept away the town, killing nearly 25,000 people (fig. 6-3). A debris flow is a mass of mud, sand, gravel, and boulders (some of enormous size) mixed with water and has the consistency of freshly made concrete. Often the mixture contains a greater volume of solid particles than of water. The word *lahar* comes from Indonesia, where it is applied to debris flows that contain mostly fragments of volcanic origin mixed with water and that are caused by volcanic eruptions. Lahars include volcanic debris flows as well as derivative floods that have a higher water content than debris flows. The people of Indonesia have experienced hundreds of lahars, and over the past four hundred years, an estimated 18,500 lives have been claimed by them.

One hundred and forty years earlier, as reported by Joaquin Acosta, a debris flow swept over the site on which Armero would later be settled and left a wide and barren alluvial plain. Everyone in the upper, narrower parts of the Lagunil-las Valley perished. In the lower part, several people fled to the heights above the flood plain and survived.

> Descending from the Lagunillas from its sources in the Nevado del Ruiz, came an immense flood of thick mud which rapidly filled the bed of the river [and] covered or swept away the trees and houses, burying men and animals. . . .
>
> On arriving at the plain . . . the current of mud divided into two branches. The much larger one followed the course of the Lagunillas toward the Magdalena; the other, after topping the high divide, traversed the Santo Domingo valley . . . and hurled itself into the Rio Sabandija, which was thus plugged by an immense dam. (Voight, "Countdown," p. 18)

Acosta later remarked in a letter that "[i]t is astonishing that none of the inhabitants of these villages, built on the solidified mud of old mass move-ments, has even suspected the origin of this vast terrain, which occupies an area at least equal to that of the province of the Rhone—" (ibid., p. 18). Acosta would have been even more astonished had he known that, 140 years later, people lived in another town on the same mud deposits. The people had for-gotten the lessons of the past, and when Nevado del Ruiz Volcano became active in 1985, government officials did not want to pay the economic or polit-ical costs of an evacuation and possible false alarm. As it turned out, in terms of the death toll, the Armero disaster was the second largest volcanic tragedy of this century and the fourth in written history—after Tambora in 1815 (92,000), Krakatau in 1883 (36,000), and Mount Pelée, Martinique, in 1902 (29,000).

Debris flows are fluids with properties completely different from those of water. The greatest distinction between a debris flow and water is viscosity, for water will flow down any incline unless it becomes frozen, but if a debris flow moves too slowly, it can become a solid regardless of temperature. A debris flow is comparable to flowing concrete—a lumpy mixture of sand, cement, and gravel that pours like a liquid down the chute of the delivery truck, but when dumped on the ground, slows down and forms a lobe with a steep front and sides. As a debris flow slows down, it reaches a threshold of viscosity beyond which it will no longer move and becomes solid. The solid is not rock-hard but, like a pile of moist sand, it does not flow easily.

Debris flows are not unique to volcanoes. For example, in the Santa Monica Mountains and other similar areas around Los Angeles, California, where de-velopers cut into the mountainsides to build pads for new houses, the slope equilibrium is disturbed, soil is used to extend the lot size, and large areas of native ground cover are stripped off. Like a volcano that has abundant new and fresh pyroclastic debris on the ridges and valleys and has no vegetation to anchor the debris, when rains soak these constructions, debris flows are created that in some cases end up in the backyards of the unfortunate neighbors below.

During wet winters following intensive construction periods or following forest fires that destroy the ground cover, tragic stories about the consequences of such debris flows often appear in the newspapers. One sleeping family was covered by a quick mudflow of such dimensions that it entered their house by pushing through the windows and doors on one side and flowed out of them on the other, leaving a mud deposit as high as the window sills. Confinement of the flow in the house below that level caused the mud to slow down and solidify. One would expect that family members rose from their beds and splashed out of the house, but because of the ability of debris flows to solidify when they stop, the family was entombed in their beds. The deposit could only be removed from inside the house by shoveling.

Volcanic eruptions can be as harmful as the developer's backhoe. Layers of volcanic ash can bury and smother great tracts of vegetation, creating a stark desert of debris that is far more extensive than any subdivision. The processes are different, but the results are the same—with heavy rains come lahars.

Lahars are initiated on volcanoes in different ways: by heavy rains falling on barren ash, by pyroclastic flows entering a river and mixing with the water, by water draining from debris avalanches, as at Mount St. Helens (chap. 1), or by eruptions occurring directly through a crater lake. They can create havoc, sometimes as highly fluid and erosive floods that cut deep gullies and carry away large objects or sometimes as debris flows that bury houses and people alike.

Moving in slow tractor-tread fashion without causing erosion, lahars can surround obstacles such as houses, cars, and people. Such slow and nonturbulent motion occurs most effectively when the amount of solids (clay, silt, sand, and gravel) is large relative to the total amount of water. Some are known to flow with only 9 percent water. Such highly concentrated, high-density flows are very viscous and can move over soft, loose items such as pine needles and other objects without disturbing them and yet carry enormous boulders for dozens of kilometers.

Lahars can move as slowly as 1.3 meters per second or as fast as 40 meters per second, and some have traveled as far as 300 kilometers. They follow valleys but their momentum may leave asymmetric water marks, the high ones being on the outer part of a bend in the river. On steep slopes they run fast and leave thin deposits, but in lowland areas they slow down and leave thick ones. Lahars momentarily dammed by a narrow constriction may form a pond several tens of meters above the valley floor that leaves a veneer of mud or "high-water" mark as the debris drains away. Mud veneers as high as 150 meters above valley floors have been reported. The lahar at Mount St. Helens that flooded the Toutle River attained a level of 2 to 3 meters above the present-day river as shown by the mudline on the trees (see chapter frontispiece).

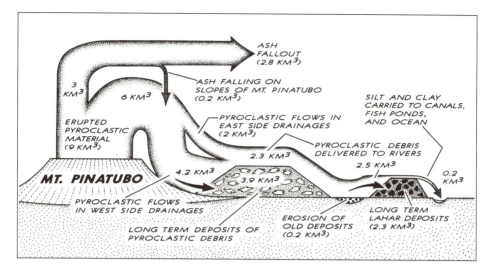

Fig. 6-4. Pyroclastic fragments are carried away from volcanoes in several ways. This diagram shows the manner of transportation and volume of pyroclastic fragments from the June 1991 eruption of Mount Pinatubo, Philippines, including ash fallout, pyroclastic flows, and water-carried debris. The thickness of the arrows and deposits is proportional to the material volume. (Modified from Pierson et al., 1992)

Destruction from lahars can continue for many years after a volcanic eruption has dumped loose ash over the landscape. Erosion, sedimentation, and flooding from the Mount St. Helens lahars cost nearly one billion dollars from 1980 to 1990 for repair and clean-up. An estimated 3.4 million kilograms of debris were washed off the slopes of the Toutle River system into the rivers during the first three years after the eruption. For nearly a decade, the Toutle River was one of the most sediment-laden rivers in the world, although that distinction has now been passed to the rivers draining the region affected by the 1991 Mount Pinatubo eruption.

In June 1991, Mount Pinatubo, in the Philippines, erupted about 12 cubic kilometers of volcanic debris as pyroclastic fallout and pyroclastic flows over thousands of square kilometers (chap. 3). As a result, Mount Pinatubo has become a gigantic storehouse of easily eroded volcanic ash (fig. 6-4). With each storm season, the ash chokes streams and rivers, causing floods and destruction of precious farmland. During the first monsoon after the eruption, damage near Mount Pinatubo resulted from burial by a lahar or flood, banks that collapsed along migrating river channels, and back-flooding from the damming of tributary valleys by volcanic deposits. In areas farther from the

Fig. 6-5. *Top*: Lahar in Sacobia Valley, near Clark Air Base, Philippines, August 4, 1991. (Photo: Thomas J. Casadevall, U.S. Geological Survey); *Bottom*: Looking east across the Central Plain of Luzon, Philippines, about 20 kilometers northwest of the summit of Mount Pinatubo. Lahars from the eruption of Mount Pinatubo (smooth surfaces, light-color) have cut across and dammed a stream, forming a lake that has flooded several farms. February 28, 1993. (Photo: Grant Heiken)

volcano, channels, drains, fish ponds, and coastal zones were filled with ash. Most bridges and roads in the lahar hazard zone were destroyed (fig. 6-5).

Many communities along the rivers draining Mount Pinatubo were flooded or buried with mud, and 86,000 hectares of agricultural land and fish ponds were destroyed, forcing more than half a million people, mostly farmers, out of work. During every monsoon season since the June 1991 eruption, lahars have been a problem and will continue to be so for many years. A 1992 study predicted that 40 to 50 percent of the newly deposited ash, together with older eroded soils and rock, will be carried away during the subsequent ten years. Damaging floods have also occurred each year as of this writing. As much as 3.6 billion cubic meters of sediment will continue to wash off the mountain before preeruption conditions begin to return.

Potential Destruction from Mount Rainier

About 5,700 years ago, a series of lahars known as the Osceola Mudflow originated from the summit and northeast flank of Mount Rainier, Washington (figs. 6-6 and 6-7). The mudflow started when part of the mountain collapsed and formed a large avalanche that cut into Mount Rainier's core, which was largely altered to clay by rising acid vapors from the magma beneath the volcano. The landslide incorporated rock debris, glacial ice, and stream water, and was transformed into a lahar as it traveled down the slopes. It moved down the White, Green, and Puyallup River drainages, then spilled into the Puget Sound lowland for a distance of over 110 kilometers. It covered a minimum area of 505 square kilometers and filled in some of the embayments on Puget Sound. The present-day towns of Buckley, Enumclaw, Pacific, Auburn, and Sumner are built on the deposits, and in other drainages from Mount Rainier, there are many more towns within reach of future lahars. Tens of thousands of people now inhabit the region. If the Osceola lahar traveled as fast as the Armero lahar (40 kilometers per hour), it could have been emplaced within two to three hours. This is hardly enough time to warn or evacuate thousands of people in the region of impending disaster should there be another lahar. Moreover, lahars follow valleys, which are where most people construct their houses, farms, and cities. People who live in the valleys draining Mount Rainier or other Cascade volcanoes need to be aware of possible dangers and to be receptive to warnings of danger for Mount Rainier still has the potential to create another large debris flow.

Yet the following advice given by Rocky Crandell of the USGS in 1971 has been ignored to this day.

> If future eruptions of Mount Rainier were to be similar in scale and type to those of the last 10,000 years, the greatest hazard would be that of lahars. Because of the

Fig. 6-6. One hundred kilometers away from Seattle, Mount Rainier appears dwarfed by Seattle's Space Needle, but the summit of Mount Rainier stands 4,392 meters higher than the cities of Seattle and Tacoma. Mount Rainier was selected as the United States' Decade Volcano for the International Decade of Natural Disaster Reduction (1990–2000) because of its recent history of activity and its proximity to the large cities of Seattle and Tacoma, both of which sit on prehistoric deposits from a mudflow that swept all the way from the volcano through the present city locales in prehistoric time. (Photo: Seattle-King County News Bureau—James Bell)

restriction of lahars to the lower parts of valleys away from the immediate flanks of the volcano, valley floors should be evacuated immediately within a radius of at least 25 miles from the volcano if an eruption should begin. It is proposed that permanent residences should not be constructed on certain valley floors near Mount Rainier, that consideration be given to the relocation of campgrounds that are now in potentially hazardous areas, and that future highways and bridges be designed and located to minimize destruction by future lahars. Likewise, the planning of all other residential, economic, and recreational developments within valleys that head on the volcano should be concerned with lahars as potential geologic hazards.

Artificial traps that might prevent large lahars from entering densely populated areas now exist in the form of hydroelectric power dams and flood-control dams in some valleys. To control a lahar, reservoirs behind these dams would have to be

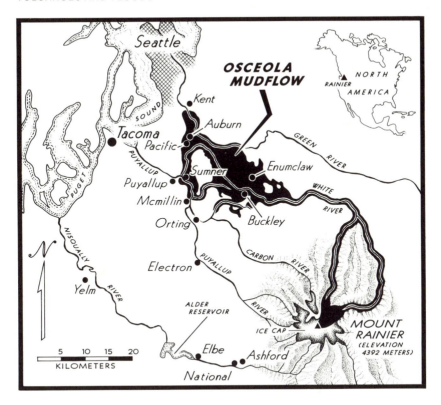

Fig. 6-7. The Osceola Mudflow, an enormous flow of debris from Mount Rainier about 5,700 years ago, traveled over 110 kilometers down the White River valley and spilled into the Green and Puyallup River valleys. The area is now occupied by hundreds of thousands of people. Seattle and Tacoma are in areas invaded by even larger and older debris flows, and the entire region, which contains millions of people, is vulnerable to possible future floods of debris from Mount Rainier.

empty. No reservoir, however, would be of any avail in controlling or diverting a lahar comparable in size with the largest that originated at Mount Rainier in postglacial time. (Crandell, *Postglacial lahars*, p. 1)

LAHARS and lava flows are well publicized geological hazards because people can stand close and watch or take photographs of them. Lahars are quite dangerous because they travel so rapidly, whereas lava usually flows slowly enough to evade, even if you are on foot. The next chapter discusses how to stop or divert lava flows.

References

Crandell, D. R. *Postglacial Lahars from Mount Rainier Volcano, Washington.* U.S. Geological Survey, Professional Paper 677. Washington, D.C., 1971.

Dragovich, J. D., P. T. Pringle, and T. J. Walsh. "Extent and geometry of the mid-Holocene Osceola Mudflow in the Puget Lowland—Implications for Holocene sedimentation and paleogeography." *Washington Geology* 22 (1994): 3–26.

Foxworthy, B. L., and M. Hill. *Volcanic Eruptions of 1980 at Mount St. Helens. The First One Hundred Days.* U.S. Geological Survey Professional Paper 1249. Washington, D.C., 1982.

Inoue Yasushi. "Under the Shadow of Mt. Bandai." In *The Showa Anthology 2: Modern Japanese Short Stories: 1961–1984,* edited by Van C. Gessel and Tomone Matsumoto, 232–54. Tokyo: Kodansha International, 1985.

Pierson, T. C., R. J. Janda, F. V. Umbal, and A. S. Daag. *Immediate and Long-Term Hazards from Lahars and Excess Sedimentation in Rivers Draining Mt. Pinatubo, Philippines.* U.S. Geological Survey Water-Resources Investigations Report 92-4039. Vancouver, Wash., 1992.

Pringle, Patrick T. *Roadside Geology of Mount St. Helens National Volcanic Monument and Vicinity.* Washington Department of Natural Resources Circular 88. 1993.

Sekiya, S., and Y. Kikuchi. "The eruption of Bandai-san." *Journal of the College of Science, Imperial University, Japan* 3 (1890): 91–172.

Siebert, L., H. Glicken, and T. Ui. "Volcanic hazards from Bezymianny- and Bandai-type eruptions." *Bulletin of Volcanology* 49 (1987): 435–59.

Sundell, K. A. "The Castle Rocks Chaos: A Gigantic Eocene Landslide-debris flow within the southeastern Absaroka Range, Wyoming." Ph.D. diss., University of California, Santa Barbara, 1985.

Voight, Barry. "Countdown to Catastrophe." *Earth and Mineral Sciences* 57, no. 2 (1988): 17–30.

Lava Flows

Frontispiece. In October 1987, the Royal Gardens subdivision, developed on the southeastern flanks of Kilauea Volcano, Hawaii, was overrun and burned by lava flows from eruptions along the volcano's east rift. These cars, parked near houses in the development, were carried along and partly engulfed by smooth, ropy pahoehoe lavas. It is usually rather easy to escape from lava flows, therefore no one died during this volcanic event. (Photo: Jim D. Griggs, Hawaiian Volcano Observatory)

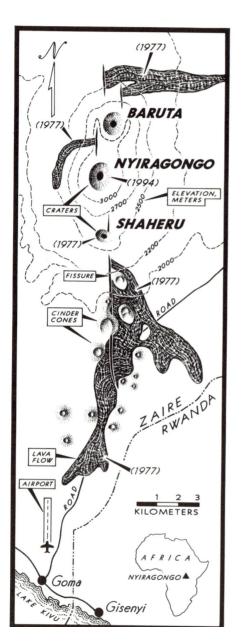

Fig. 7-1. Nyiragongo Volcano, Zaire, with craters, cinder cones, fissure vents, and lava flows erupted in 1977 and 1981. Very fluid and consequently very fast lavas erupted from flank vents on January 10, 1977, and swiftly engulfed homes and villages, killing seventy people. Recent concerns about this frequently active volcano have been raised because Goma, the site of Rwandan refugee camps, lies on its slopes. (Adapted from M. Krafft, *Global Volcanism Newsletter*, and the Smithsonian Institution)

From about 1928 to 1977, Nyiragongo Volcano had filled with a fiery lava lake in a crater about 1.5 kilometers in diameter at 3,469 meters above sea level. From time to time the lake upwelled and overturned, bringing the hottest and reddest lava to the surface, which then cooled, darkened, and sank, then remelted and rose again. The brew splashed and hissed, the heat made the air shimmer, and gases fountained like geysers to make high incandescent lava streamers. During this time, the level of the lake changed continuously, but on the average, it remained high. Then, on January 10, 1977, a fissure opened and the lava suddenly emptied from the crater. It poured out onto the surrounding countryside, moving at an astounding initial velocity of 100 kilometers per hour, and overwhelmed everything in its path. Seventy people were killed and eight hundred lost their houses to the highly fluid lava.

Nyiragongo Volcano is one of two active volcanoes in Zaire's Virunga Range, which comprises eight large volcanoes (fig. 7-1). Nyiragongo is perpetually active, and built of basaltic lava flows interspersed with some pyroclastic layers. Before the lava drained out, lava flows from Nyiragongo had never been seen outside its crater, although there had been thirteen eruptions between 1894 and 1977.

On January 10, a set of fissures opened north and south of the summit at elevations of 2,000 meters and 1,500 meters on the lower flanks of the volcano, and the lava emptied from the crater. An estimated 20–22 million cubic meters of lava drained in half an hour and flowed for 20 kilometers. The lava's hydraulic heads (1,800 meters and 1,300 meters), its temperature (about 1000°C), and its low viscosity (less than 40 percent silica) account for the speed with which the lake emptied and for the lava's high velocity. With an average speed of 30 kilometers per hour, the front of the lava flow reached the outskirts of the village of Goma (which lies on the shores of Lake Kivu on the edge of the Virunga Volcanic Field) in twenty minutes and stopped 600 meters from the Goma airport. The extensive surface area of the very hot lava generated extreme temperature gradients, which in turn caused high winds that uprooted large eucalyptus trees and banana groves.

The 100 kilometers per hour estimated for the initial speed of the Nyiragongo lava is the fastest reported for a lava flow, and so it was able to surprise all of the living creatures in its path. Other fast flows have been recorded in Hawaii—a speed of 64 kilometers per hour was reported for an 1855 lava flow of Mauna Loa that was moving down slopes of 10 to 25 degrees; in 1940, flow

Fig. 7-2. The effects of volcanoes can invade every aspect of life, and even death. Lava flows from the 1959–1960 east rift eruption of Kilauea Volcano, Hawaii, partly buried a Hawaiian Buddhist cemetery near the village of Kapoho. (Photo: Richard V. Fisher)

Table 7-1. Attempts to Conquer Lava Flows

1669	Stone walls built to protect the city of Catania, Sicily—Failed.
1960	Levees built to divert aa flows during the Kapoho eruption, Hawaii. Water jets from fire trucks used to delay advance of lava flow at selected areas of the village of Kapoho—Failed.
1973	Water jets from harbor boats aimed at flow front threatening to close the entrance to the fishing harbor—Successful.
1975, 1976	Bombing of lava tunnels and levees from aircraft to divert flows from Mauna Loa, Hawaii—Eruption stopped.
1983	Earthen barriers constructed and lava diversion at Etna, Sicily—Partial success.
1986	Water jets used on the front of a lava flow that advanced to within 200 meters from the town of Motomaki, Japan, on the island of Izu-Oshima—Successful.
1991–1992	Multivaried tunneling, explosive blasting, artificial levees, wire mesh, and concrete plugs in lava tunnels at Etna, Sicily—Partially successful.

velocities of Mauna Loa lavas reached 30 to 40 kilometers per hour. The bigger and faster the flow, the farther it can travel. The one at Nyiragongo in 1977 flowed 20 kilometers, and the one from Mauna Loa in 1859 covered 50 kilometers. Kilauea flows are rarely more than eleven kilometers long. But most lava flows move rather slowly, and the loss of life from them is a small fraction of that from other volcanic phenomena, such as lahars, pyroclastic flows, and avalanches. Most lava flows are usually slow enough to allow everyone to escape, and some flows creep so slowly that a person can stand in front of one to collect samples without being engulfed.

A slow-moving lava flow is a behemoth that pushes or moves over everything in its path. In some cases entire villages, including cemeteries, have been pushed over and buried under solid rock (see chap. 11) (fig. 7-2). It is next to impossible to recover buried property that is covered by lava, and no matter what the lava thickness, agricultural lands cannot be cultivated again because the ground is blanketed by solid rock for decades.

Stopping a Lava Flow

During the last three hundred years, there have been some attempts to stop or turn aside slow-moving lava flows, usually with limited success. Various methods include using natural or man-made diversion barriers, cooling the lava with water, or interrupting the supply of lava by bombing and blocking the lava tunnels (table 7-1).

Ditches, Walls, and Dams

Lava usually moves slowly enough that a ditch or a high wall may change the flow's direction. There have been many examples of lava diverted by small hills, gullies, and barriers built for other purposes—such as stone walls, railroad embankments, and levees—as well as by walls built specifically for the purpose of diversion. Yet, there have been an equal number of cases where slow-moving lava merely piled up higher and higher until it overtopped the obstacle and continued in the same direction as before. But successful examples of diversions have inspired the idea of using artificial barriers to direct lava flows into less harmful paths.

In Hawaii in 1955, small debris dikes were piled up by bulldozers to divert a lava flow from Kilauea, but success was limited. Another, more successful attempt was made in 1960 near the town of Kapoho. The height of a natural rise in front of a lava flow was increased to save a lighthouse, and other barriers were constructed to divert the flow. Bulldozers erected six dikes, as long as 2,295 meters, from 1.5 to 7.6 meters high, and from 22 to 27 meters wide. The lava moved inexorably over the dikes, which caused a slight change in direction as the lava approached the lighthouse, and it was saved. From his experience with attempts to divert lava flows in 1955 and 1960 in Hawaii, Gordon Macdonald, of the University of Hawaii, said that diversion barriers may work in a limited way for very fluid and thin pahoehoe lavas but are generally ineffective for the more viscous aa or block basalt lava flows. Macdonald also said that a barrier needs a very wide base and must be composed of denser material than the liquid lava, otherwise the lava could break through and carry the barrier away.

Many of Macdonald's conclusions, however, were refuted at Mount Etna in 1983. Four earthen barriers, ranging in size from 300–580 meters long and 8–20 meters high, channeled enough aa lava to save important installations, including an astrophysics observatory. The barriers were built of low-density scoria, lapilli, and ash but were not carried away by the denser lavas. The purpose was not to stop a lava flow but to divert it toward unoccupied zones around the volcano. But a barrier built at Portella Calanna on the flanks of Mount Etna in January 1992 was meant to actually dam the lava and slow it down. The dam was 21 meters high and was placed at right angles to the flow's direction. It was also built of low-density scoria, lapilli, and ash, but even so it withstood the pressures of a lava flow for a month and diverted it into the large Val Calanna basin. The flow piled up against the dam, accumulating thin flows of fluid lava that broke out from the front of aa flows. Although damming also runs counter to Macdonald's advice, its effectiveness depends upon local terrain. This dam was built at the narrow outlet of a large valley that provided a large upstream basin to contain the lava and slow down the flow's advance. It was found that the type of construction material and shape of the structure is less important than height of the dam. Also, the dam was shaped to encourage lateral spreading and reduce erosion during overflow. The purpose of the dam, like other types of intervention, was to gain time, because any embankment will be overrun by lava if rate of flow is high for a long enough period of time.

Cooling Lava with Water

At Heimaey, Iceland, in 1973, a lava flow threatened to overwhelm the town of Westmannaeyjar and to completely close off one of the most important harbors in Iceland, thereby threatening the fishing industry. The lava was stopped by cooling it with water jets, and the saga of how it was done is dramatically chronicled by John McPhee in *The Control of Nature*. Such a tactic had its doubters, but the believers held their ground and poured massive amounts of water from the nearby sea onto the advancing flow. With this unlimited supply, jets of water from boats and from land were pumped onto the flow's front and flanks at a rate of 900 liters per second. In the spots where large volumes of water could be sprayed onto the flow, there was great success at stopping the lava, although diversions arising from natural causes were difficult to distinguish from those of the pumped water. Some people believe that it stopped because the volume of lava decreased to such an extent that it could no longer advance. That was true at various places, but there were clear successes at modifying flow speed and direction. Whether fortuitous or not, the flow benefited the community, for it decreased the width of harbor entrance and thereby increased the protection of the harbor from the strong waves of the North Atlantic.

Pouring water on a lava flow to stop it had been first tried in Hawaii. When Kilauea erupted in 1960, lava moved through the town of Kapoho, destroying one house after another. The fire department tried to slow the lava by cooling it with jets of water and managed to control the expansion of the flow margin for a few hours, which gave the residents time to remove some possessions from their houses.

On Mount Etna in 1983, lava was prevented from laterally overflowing from its channel when the fire brigade sprinkled the lava with jets of water from their trucks. The modest quantity of water formed a solid crust that was sufficient to retain the lava in the channel for a few hours and keep the lava from flowing into a yard. A wall of solid lava grew above the previous dike, but the lava flowed into the yard when the water supply was depleted.

Water jets are useful in some cases if they are directed at the flow's front and flanks, making it easier for the lava to advance in another direction. Spraying only the front is useless and even dangerous because it may favor tunneling of the lava, causing it to break out unexpectedly. Stopping lava by cooling it requires extraordinarily large amounts of water to achieve success because of the great heat capacity of fluid lava.

Interrupting a Lava Flow

One of the most effective lava-fighting methods is to halt or reduce the upstream supply of lava. Upstream reduction of the flow limits the invasion of new downstream areas, which in turn reduces problems that result when lava is diverted from one place only to cause damage at another. A case in point is the 1669 Catania lava flow from Mount Etna. Workers armed with picks and shovels attempted to break the crust on one side of the flow and cause it to move away from Catania. Had the flow moved in that direction, the village of Paterno would have been at risk, and to forestall that possibility, five hundred enraged citizens of Paterno forced the lava fighters to flee. As a consequence of the dispute, a royal decree was issued stating that, henceforth, no one was to interfere with the natural path of a lava flow. Almost nothing could have been done, even in modern times, to prevent the damage caused by the that particular flow because the lava poured out so fast and traveled so far. The vents had opened at 800 meters elevation near the town of Nicolosi, and within twenty hours, the lava had reached and destroyed another village just

3.5 kilometers farther downhill. Other villages were destroyed during the following days, and thirty-three days after the flow began, flows reached the city walls of Catania.

When a lava flow moves as a river, its upper part may crust over to form the roof of a tube that insulates the hot lava within, thereby conserving heat; the lava can therefore remain liquid longer and flow much farther. Stopping a lava flow was first tried in Hawaii in 1935 by bombing the roof of a lava tube in a flow from Mauna Loa. The vent was at an elevation of 2,743 meters, but far below at sea level, the flow front was threatening the port town of Hilo. The tube was blown apart and became clogged with solid chunks of lava, but the effectiveness of the bombing was never determined because the eruption stopped. In 1942, another bombing attempt was made to breach a flow and cause side flows so as to reduce the volume of the main flow. The bombing had no significant effect, but Hawaiians were upset because the bombing was an affront to Madam Pele. The tried-and-true way to stop lava flows is to make offerings of alcohol, flowers, food, tobacco, or nearly anything else to appease the goddess.

Irrespective of Madame Pele's feelings, in 1975 and 1976 experiments were conducted that had been designed to learn more about the effectiveness of bombings in stopping lava flows. Bombs weighing up to 900 kilograms were dropped on a cinder cone, on the flank of a lava channel, and on the roof of a lava tube produced by a prehistoric eruption of Mauna Loa. The roof of the tube was barely affected, the flank of the channel was destroyed only where it was thin, and the bombs only created wide craters in the cinder cone.

The 1991–1992 Eruption of Mount Etna

Mount Etna began erupting on December 14, 1991 (fig. 7-3), when two fractures, one trending north and the other southeast, developed at the base of Southeast Crater near the summit. Ash and lava fountains began and lava moved downslope. The most extensive flow erupted during the night of December 14 and 15 at the southernmost end of the southeast fracture on the western wall of Valle del Bove. The initial extrusion rate was 18 to 25 cubic meters per second, but it then slowed to 10 to 15 cubic meters per second. During the ten and one-half months of activity, over 500 million cubic meters of lava poured out of the fissure, making it the largest eruption in more than three hundred years. It was second only to the 1669 eruption, which emitted more than 900 million cubic meters of lava that traveled 17 kilometers to the sea, destroying dozens of villages and the western part of the city of Catania.

When the 1991 flow began, a computer simulation of the most probable paths for the lava was conducted, and it was predicted that the village of Zafferana Etnea would be destroyed if the rate and volume of lava flows were

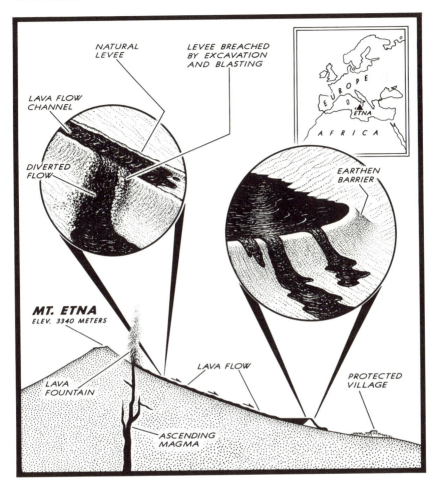

Fig. 7-3. Mount Etna, Sicily, has been erupting since historic time, but mankind's en-croachment has resulted in many encounters with the volcano even though the eruptions produce slow-moving basaltic lava flows. Abundant historic cinder and spatter cones that mark the vents are mostly high on the mountain and do not threaten the towns and cities on the lower slopes, but lava flows have been a constant threat throughout time. Sicilians have tried for hundreds of years to control lava flows, but only recently, during the eruption of 1991–1992, have they had some success with modern machines. Earthen barriers (inset) slowed the lava flows but did not stop them. The most effective means of control was to divert lava flows near the source, high on the mountain, by breaching the natural lava levees (inset). (Adapted from Barberi et al., 1993)

large enough. Following the initial outpouring of lava at 25 cubic meters per second, the simulation indicated that the flow could travel as far as 20 kilometers and overwhelm Zafferana Etnea on the way.

By March 14, 1992, the flow front had moved to within 2 kilometers of Zafferana Etnea, and from then until May 27, 240 million cubic meters of lava were emitted. As the eruption continued, the lava flow easily formed tunnels at distances of over 8 kilometers from the vent. Roofing of the lava flow reduced the amount of heat loss from the flow, which therefore could maintain its high temperature and fluidity and ability to flow farther. Despite the interventions upstream from Zefferana (see above), the lava front advanced to within 700 meters of the village and stopped. The village had been saved, but upstream from the village, the houses, roads, and orchards of Val Calanna in the zone of Piano dell'Acqua were devastated, and the water collection system of Zafferana was destroyed. Total damages were estimated at about $3 million (U.S.).

Lava Put to Use

Live lava flows may be destructive to the environment, but ancient solidified flows are quite useful to mankind. Throughout the centuries, Europeans have utilized many types of rocks for buildings, walls, and cobblestone streets. One of the most versatile type of rock is columnar-jointed lava, which can be used for decoration, monuments, walls, barriers, and street pavement, as well as a foothold for recreational rock climbing (fig. 7-4). When cut, columns can be stacked like cordwood to build nearly impenetrable walls (fig. 7-5). In Holland, the Dutch have utilized basalt columns from volcanic areas of Europe to build extremely resistant dikes that keep the waters of the North Sea from flowing over the agricultural land that has been recovered from the sea.

Lava columns are formed as the lava cools. Because molten lava is less dense than it is in its solid state, when the lava cools and solidifies, it shrinks and forms cracks called *joints*. When a lava crust is only 4 or 5 mm thick,

Fig. 7-4. Novice rock climbers in central France using columnar basalts for learning their craft. Vertical joints in the rock intersect horizontal joints as the result of slow cooling of the lava flow. The climbers are at the base of the central column, and their rope dangles from the top of the column. (Photo: Richard V. Fisher)

Fig. 7-5. Blocks of columnar, jointed lava stacked horizontally will fit together with little or no mortar because of their polygonal five- to six-sided shape. This photograph is part of a 15-meter-high wall that supports a railroad track across low ground like a bridge. The largest columns in diameter are 50 centimeters across. Many dikes in Holland are similarly constructed, as are many village "cobblestone" streets throughout Europe. (Photo: Richard V. Fisher)

Fig. 7-4

Fig. 7-5

cracks form in it and begin to define five- and six-sided shapes; as the crust thickens, the cracks continue to grow longer, carving columns from the lava. When the lava pool is completely solidified, the cracks extend from the top to the bottom of the lava flow, sometimes with columns exceeding 30 meters. Each column is a polygonal body that fits other five- or six-sided neighbors like a jigsaw puzzle. When stacked horizontally, the fitted columns form an impregnable barrier and can support extremely heavy weights. Little mortar is necessary because of the close fit of the columns.

A Ride on a Lava Flow

In Kamchatka, Russia, in 1938, a volcano called Biliukai began to erupt voluminous lava flows, and within six weeks, had built a cinder cone 109 meters high. During the eruption, the maximum production of lava had reached about 300 cubic meters per second (73,333 gallons per second). Biliukai's activity was being studied by two volcanologists, V. F. Popkov and I. Z. Ivanov. One of their projects was to determine the temperatures within the flow and the type of gases being released from it, and while doing so in November 1938, they completed and survived one of the most foolhardy adventures in the annals of volcanology:

> About 650 yards from the seepage site, the surface of the lava stream was covered by a dark crust. The flow was separated from the banks of the "river" by a seven foot wide zone of incandescent lava. The cracked crust was about one foot thick, and at points of surface unevenness almost 20 inches high. The flow bore it at a speed decreasing downstream, of 140 to 100 feet per minute.

> It was very simple to drive a steel bar into the reddish lava along the bank, and to insert a pyrometer point. However, we were forced to abandon this technique, for the moving flow would likely damage the instrument. Trying to follow the flow, holding the pyrometer by hand and, in addition, the connected galvanometer, was just about impossible in view of the heat and the ruggedness of the bank.

> Nonetheless, we were tempted to take lava temperatures. We also wanted to take samples of the gases present in the molten basalts. The idea occurred to us to cross the red-hot lava zone to get a footing on the dark crust.

> A heavy block of cooled rock which we threw upon the solidified lava surface convinced us that the latter was sufficient to hold the weight of a man. We bombarded the solidified mass of basalt which we planned to reach and stand on with sizable pieces of rock to test its resistance; it appeared solid in spite of the sound of broken glass produced by our missiles.

> Following such preliminary "tests" and without letting go of Ivanov's hand, I put—after taking infinite precautions—one asbestos-shod foot on the incandescent lava. Noting that its surface offered sufficient resistance, I released Ivanonov's hand and made another step by resting my body on the iron rod which I used as a walking stick and which sank slowly into the plastic mass.

Another step, and I gained a footing on the dark crust of floating lava where my chemist soon joined me with his equipment. We drifted along at just about the same speed as the molten lava, and we were able to measure its temperature and take samples of gas at several points.

After recording the hour and the site of the beginning of our experiment, we started to work. With a steel rod, we pierced the reddish lava to a depth of approximately 16 inches, inserted the pyrometer point and connected it to the galvanometer. The needle rapidly advanced on the indicator to the calibration corresponding to 400 degrees Centigrade, then, progressing evenly though at a gradually decreasing rate, it successively reached 500, 600, 700 and 800 degrees Centigrade. It finally came to rest at 870 degrees and then began to oscillate downward by 10 to 15 degrees Centigrade. A subsequent experiment produced a maximum of 860 degrees.

The moving flow proceeded ceaselessly but smoothly in an east-northeast direction. As we slowly drew away from the seepage site, the number of bubbles bursting at the surface grew smaller. Finally, we began to take gas samplings. The first attempts to do so failed, for we tried to collect the samples from the bubbles. However, it was found that when the bubbles were covered with a funnel, they burst not inside but rather on the side, allowing the gases to escape into the atmosphere. We then made a hole in the hardened lava surface, allowing us by means of a porcelain tube and an asbestos funnel of a sufficiently large diameter, to collect a few quarts of gas.

Subsequent analysis of the gaseous mixture made by Ivanov indicated: 1,000 mg of water per liter of gas; 0.5% hydrochloric acid; 21% oxygen and 78.5% nitrogen.

Taking lava samples involved certain difficulties. It was relatively easy to detach plastic shreds from the mass, but it was difficult to carry them to the surface of the dark crust on which we were standing, and once placed there the samples became adhesive before acquiring sufficient hardness to prevent it.

It was dangerous to remain standing for long periods of time on the moving and burning slab whose surface temperature reached 500 degrees Centigrade, and revealed a dark reddish color at the cracks. At regular intervals we were forced to climb the uneven points of the crust to allow the wind to cool our asbestos-shod feet. During our experiments we spread an asbestos sheet under us, but nevertheless we often had to balance one-footed like a stork to allow the other one to cool.

At 9,000 feet from the source outlet, the current moved along a slope of five to six degrees, and its speed was considerably reduced. We had remained about one hour on the dark crust. From our departure point we had traveled on and with the flow more than 6,000 feet in an east-northeast direction and now, having suffered no harm, we took a foothold on an older, already cooled flow. (Popkov, quoted in Wilcoxon, *Chains of Fire*, pp. 188–90)

DESTRUCTION by lava flows is an old-fashioned volcanic hazard—such events have been taking place nearly as long as there has been an earth. But the latter two decades of the twentieth century have brought us in contact with a totally new hazard caused by advances in modern technology. High-flying passenger jets have encountered eruption clouds of ash, causing their engines to shut down and the jetliner to fall thousands of meters. Interested jet passengers are invited to examine the next chapter.

References

Barberi, F., M. I. Carapezza, M. Valenza, and L. Villari. "The control of lava flow during the 1991–1992 eruption of Mt. Etna." *Journal of Volcanology and Geothermal Research* 56 (1993): 1–34.

Blong, R. J. *Volcanic Hazards: A Sourcebook on the Effects of Eruptions.* Sydney: Academic Press, 1984.

McPhee, John. *The Control of Nature.* New York: The Noonday Press, 1989.

Popkov, V. F. "On the activity of Bilukay in 1938–1939." *Bulletin of the Kamchatka Volcanological Station* no. 12 (1946): 29–33.

Tazieff, H. "An exceptional eruption Mt. Nyiragongo, January 10, 1977." *Bulletin of Volcanology* 40 (1977): 189–200.

Wilcoxson, K. E. *Chains of Fire: The Story of Volcanoes.* Philadelphia: Chilton Books, 1966.

WOVO [*World Organization of Volcano Observatories*] *News* no. 1 (winter 1993): 1–15.

CHAPTER **8**

Never Sail through an
Eruption Cloud

Frontispiece. The young Japanese lovers perched
on a crater rim enjoy a quiet time in the volcano's
activity. (Photo: Richard V. Fisher)

Fig. 8-1. Many volcanic eruption clouds are laced with lightning, a common phenomenon caused by the build-up of electrostatic charges due to millions upon millions of volcanic particles skimming and grazing each other, as in this 1971 eruption of Volcan Cerro Negro in Nicaragua. (Photo: José Viramonte)

A Sailing Ship and a Jet

On August 26, 1883, the *Berbice* was sailing through the Sunda Straits from New York to Batavia (now Djakarta) carrying a cargo of petroleum. As the ship entered the straits in the early evening, thunder and lightning were reported in dark clouds ahead (fig. 8-1). Captain Logan shortened all sail for the oncoming storm, expecting the ship to enter a heavy shower, but instead the rain was dry volcanic ash. The ship was sailing toward Krakatau Volcano, which had erupted as they entered the straits.

Krakatau Volcano is located near the middle of the Sunda Straits, an important gateway for ships traveling from the South Pacific and southeast Asian countries to Europe via the Suez canal.* Krakatau and the straits lie between what was then Dutch Java and Sumatra, two of the largest islands that are now part of Indonesia.

A ship's captain is trained to be a careful observer, thus we have witness to the beginning of the eruption from Captain Logan's log:

> The ash shower is becoming heavier, and is intermixed with fragments of pumice-stone. The lightning and thunder became worse and worse; the lightning flashes shot past around the ship; fire-balls continually fell on the deck and burst into sparks. We saw flashes of lightning falling quite close to us on the ship; heard fearful rumblings and explosions, sometimes upon the deck and sometimes among the rigging. The man at the wheel felt strong electrical shocks on one arm. The copper sheathing of the rudder became glowing from the electric discharges. Fiery phenomena on board the ship manifested themselves at every moment. Now and then, when any sailor complained that he had been struck, I did my best to set his mind at ease, and endeavoured to talk the idea out of his head, until I myself, holding fast at the time to some part of the rigging with one hand, and bending my head out of reach of the blinding ash shower which swept past my face, had to let go my hold, owing to a severe electric shock in the arm. I was unable to move the limb for several minutes afterwards. I now had sails nailed over the hatches lest the fire falling around should set my inflammable cargo in a blaze. I also directed the rudder to be securely fastened, ordered all the men below, and remained on deck with only my chief officer. At 2 a.m. on Monday, the 27th, the ashes, three feet thick, were lying on the ship. I

* Even the Suez Canal has relied upon volcanic eruptions. Opened in 1869, the canal was lined with high-quality concrete made with fine-grained volcanic ash quarried on the Greek island of Thera.

had continually to pull my legs out of the ashy layers to prevent them from being buried therein. I now called all hands on deck with lanterns to clear away the ashes, though the weather was unchanged, and the fearful electric phenomena, explosions, and rumbles still continued. The ashes were hot, though not perceived to be so at the moment of their falling on the skin. They burned large holes in our clothing and in the sails. At 8 in the morning there was no change. At that hour it was still quite dark, and the ash showers were becoming heavier. Clearing away the ashes was continually proceeded with until 11 a.m. A high continuous wind set in from the SE, which varying afterwards made the ship list considerably from the weight of ash on our masts and rigging. A heavy sea came rushing on about 3 in the afternoon. It rose to a height of 20 ft, swept over the ship, making her quiver from stem to stern with the shock. Meanwhile, the storm continued. The mercurial barometer did not stand still from 28 to 30 inches. When I went to examine my chronometers, I found that they all had stopped, probably owing to the shaking of the ship from the concussions. Up to 6 p.m., the darkness and the storm still continued, but the sea had become calmer. The flashes of light showed it to be covered with pumice and ash on all sides. After this the weather moderated and the sky cleared. (quoted in Simkin and Fiske, *Krakatau 1883*, pp. 101–2)

The wild fluctuations of the barometer were caused by overpressures from volcanic blasts. A great surprise to the captain and his crew was the intense electrical discharges from the ash cloud. Volcanic ash particles are nearly saturated with electrical charge, causing spectacular, continuous lightning displays and a less continuous electrical discharge called St. Elmo's fire, which lights up metallic surfaces and makes a crackling sound. Sometimes St. Elmo's fire occurs as balls of fire on sticks and masts.

Nearly one hundred years later, a jetliner carried its passengers headlong into a confrontation with another Indonesian volcano. This time it was Galunggung. There were the same sensations as a century ago—the smell of sulfur, St. Elmo's fire, and damage from ash—and the modern ship nearly crashed as it passed over the island of Java, not far from the Sunda Straits, where Captain Logan described his terrifying night (fig. 8-2). It was on June 24, 1982, that British Airways Flight 9, a Boeing 747-200 with 247 passengers and 16 crew members, was flying at an elevation of 11,470 meters from Kuala Lumpur, Malaysia, to Perth, Australia. Dinner had been served and night had settled as the plane crossed southern Sumatra and western Java. Minutes earlier it passed over the Sunda Straits and Krakatau. The flight had been uneventful until Captain Eric Moody left his seat to check on the main cabin. He had barely reached the bottom of the stairs when he was called back to the flight deck. Running up the stairs he saw the flight engineer and co-pilot watching a spectacular display of St. Elmo's fire out of the control-room windows. It was so intense that it looked as if magnesium flares were in the engines. Then, a series of impossible events occurred. First, the number four engine failed. It

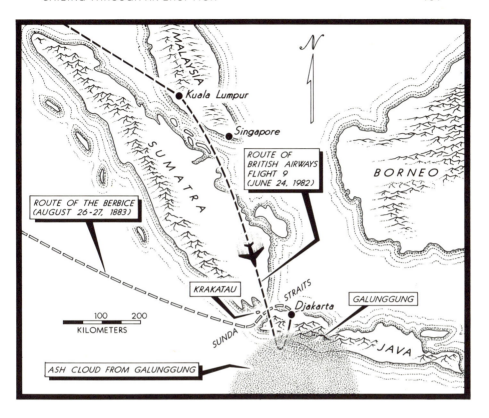

Fig. 8-2. The vessel *Berbice*, sailing from New York to Batavia, endured and survived the 1883 eruption of Krakatau. Krakatau is located in the Sunda Straits on an important shipping route. Because of his knowledge of natural phenomena and familiarity with his ship, Captain Logan led his crew safely past one of history's largest volcanic eruptions. Nearly one hundred years later, British Airways Flight 9, from Kuala Lumpur to Perth, also flew over the Sunda Straits. A few minutes later, the Boeing 747 flew through the periphery of a volcanic ash plume from the eruption of Galunggung Volcano on Java. Volcanic ash sucked through the engines caused all four of them to shut down. The plane fell powerless, but at the last minute, Captain Eric Moody and his crew were able to restart the engines and made an emergency landing at Djakarta (formerly Batavia).

shut down over the next thirty seconds. Then, one after the other, they all failed. With great reserve, the flight engineer said, "Number two's gone, number three's gone, and . . . golly-gosh, we've lost the lot."

Four engines on modern jets do not fail; it simply does not happen. Mystified, the crew sent an immediate mayday call. "Djakarta, Djakarta, Mayday, Mayday. This is Speedbird 9. We have lost all four engines. Repeat, we have lost all four engines! . . . There is a possibility that we may have to ditch." The radio transmission was difficult to understand because of the tremendous static from electrical discharges. They thought they had heard right, but air traffic controllers in Djakarta simply did not believe it possible that four engines could have failed at the same time.

Beginning at an elevation of 11,470 meters, the 747 became an enormous glider as the crew tried again and again to restart the engines. During the first check they fell as much as 930 meters and then ran through the restart procedures at least twenty times after that. At 4,030 meters, one engine was restarted and another started about a minute and a half later. Twenty seconds later, the remaining two engines came on with an enormous roar. But the number two engine surged badly, causing the plane to lurch from side to side, so Captain Moody ordered it shut down in fear that it might shake itself off of the wing.

As they turned into the approach to Djakarta airport for an emergency landing, they realized that the visibility was extremely poor because the windows had been sandblasted, and the crew could only see rather poorly through about a two-inch strip on either side of the windshield. The captain had to stand, peering through the side of the window, flying the plane on three engines. At an elevation of about 30 meters, Captain Moody remarked "Oh well, we aren't going to die now," and the plane made a smooth landing.

Unlike Captain Logan's experience one hundred years earlier, when he knew that it was ash and pumice showering onto the deck of his ship, the British Airways crew had no inkling that they had flown through a volcanic eruption plume. Since it was night when the jet passed through the eruption cloud, the crew could not see it, but even during the day, a disseminated eruption plume doesn't look much different from an ordinary cloud made up of condensed water vapor. Nor are eruption plumes dense enough to be visible on present onboard radar systems. In 1982 there were no warning systems nor any awareness that they were needed. Galunggung, located in south central Java, had been erupting for three months, with ash plumes sweeping east and south. Thousands of residents had been evacuated, but not a single thought was given to flights passing overhead.

Unlike the crew of the *Berbice*, who had hours to evaluate their situation, the crew of British Airways 9 had only minutes to keep the plane in the air. It was only the calm, highly professional actions of the pilots that prevented disaster. Ship and flight captains must have an understanding of natural hazards that could endanger them and their passengers. These hazards include storms and ice build-up and less well known ones, such as ash plumes from volcanic eruptions. In the case of these two eruptions in Indonesia, the captains and their passengers survived because they knew their craft well and not because they understood the hazard. However, with increased communications, prediction, and education, modern captains can better deal with mitigation of the hazard.

Other Encounters with Ash Clouds

Another encounter of a commercial aircraft with an ash cloud occurred on December 15, 1989, about 240 kilometers northeast, and completely out of sight of, the erupting Redoubt Volcano, Alaska, as a plane was descending for a landing in Anchorage (fig. 8-3).

Redoubt Volcano is 177 kilometers southwest of Anchorage, Alaska's largest city and a major hub of domestic and international commercial air travel. Redoubt is a 3,108-meter-high andesite composite volcano near the end of the Aleutian volcanic arc and sits above the Aleutian subduction zone, where oceanic crust of the Pacific plate rides beneath the North American plate. This volcano has erupted at least thirty times over the past ten thousand years, its oldest historic eruption being in 1902. Vapor emission occurred in 1933 and then between 1965 and 1968. Strong explosive activity produced eruption columns of ash and flooding in January 1966, and a lava dome was emplaced in the summit crater sometime during 1966–1967.

The early hydrovolcanic and magmatic explosive activity of Redoubt Volcano that occurred December 14–16, 1989, produced several ash-laden eruption columns. The most voluminous ash production and highest column from Redoubt occurred during the major explosive event on December 15. Strong upper-level winds carried the ash into interior Alaska and northwestern Canada, more than 500 kilometers from the eruption. However, unlike what happened in Indonesia in 1982, by 1989, warning systems were in place and all airlines had been notified of the eruption. Despite the warning, a new KLM Boeing 747-400 jetliner, enroute from Amsterdam to Tokyo via Anchorage, with 231 passengers and 14 crew members, flew through an ash plume east of Talkeetna, Alaska, at an altitude of 7.5 kilometers on December 15 at 11:50 A.M. Suddenly, all four engines shut down. For twelve minutes the jet steeply glided downward for 4.0 kilometers before the crew managed to restart the engines—1,500 meters above the mountain tops. It then made a safe landing at Anchorage International Airport, but the damage to the engines, avionics (electronic devices), and the aircraft's structure cost more than $80 million. Subsequently, four other commercial jetliners suffered damage from the Redoubt eruption cloud on December 15 and 16 and on February 21, 1990, but their engines did not fail.

Heavy ashfall downwind of Redoubt Volcano severely disrupted air traffic above southern Alaska. Many domestic carriers suspended service to Alaska, canceling hundreds of flights and causing long delays at airports on the West Coast and in Alaska over the Christmas holidays. Several international carriers continued to reroute flights around Alaska through January 1990. Military aircraft canceled some flights and rerouted others. In all, canceled flights caused an estimated loss of $2.6 million to Anchorage International Airport. Further economic disruptions came from the threatened Drift River Oil Terminal near

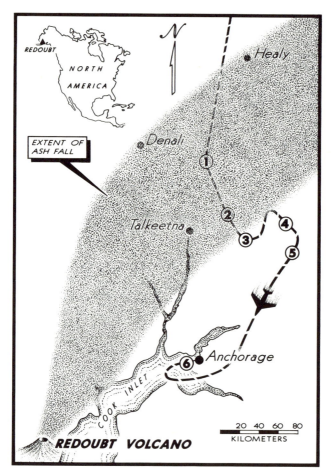

Fig. 8-3. The 1989 eruption of Redoubt Volcano, Alaska, produced eruption clouds that were swept northeasterly by the wind. On December 15, 1989, a KLM 747-400, flying from Amsterdam to Tokyo with a stop in Anchorage, encountered an ash cloud from Redoubt. (1) 11:40 (local time)—the flight began a descent from 10.7 kilometers and changed course to avoid the cloud; (2) 11:47—the ash cloud was encountered at 7.5 kilometers; (3) 11:47—power on all four engines was lost when an attempt was made to climb to 8.5 kilometers; (4) 11:52—engines 1 and 2 were restarted at 5.2 kilometers; (5) 11:55—engines 3 and 4 were restarted; (6) 12:25—the plane landed safely at Anchorage International Airport. Repair costs to the plane were about $80 million. (Adapted from Casadevall, 1994)

the volcano that resulted in the shut-down of Cook Inlet oil production, and a consequent decrease in the overall production of goods and services in Alaska.

Beginning December 13, the Alaska Volcano Observatory (AVO) had been monitoring Redoubt Volcano twenty-four hours a day with seismic instruments and slow-scan TV systems on the ground and from fixed-wing aircraft and helicopters. Whenever there were strong seismic tremors for more than a few minutes, a large, vigorous eruption plume occurred. This information was relayed within minutes of an eruption's onset to the Alaska Division of Emergency Services, to the FAA, and to the Drift River Oil Terminal. The FAA would then relay the information to commercial airlines who relayed the information to their pilots. Redoubt is just one of many active Alaskan volcanoes for which the AVO provides vital warnings to air traffic over thousands of square kilometers.

Jetliners

Millions of people ride jetliners every year, but few passengers are aware of the hazards of volcanic eruption clouds. Many air travelers have experienced turbulence and excessively bumpy rides, but fortunately, relatively few people have been on a commercial jet when all of the engines have failed at the same time. There have been about eighty incidents in which turbine aircraft have suddenly encountered eruption clouds. Several aircraft were damaged from flying through the ash clouds of the 1980 eruption of Mount St. Helens, Washington.

The encounters of aircraft with ash have cost tens of millions of dollars in each case, but no lives have been lost. Ironically, since 1982, the increase in engine failures due to ash ingestion results from the increased efficiency of jet engines. They consume less fuel and run at much higher temperatures than older aircraft. At cruising altitudes, 7.5 tons of air per minute passes through each of the four engines on a Boeing 747—that is equivalent to a volume of 18,000 cubic meters. The engines can quickly filter out substantial amounts of volcanic ash if it is present in this great volume of air. If there are 250 milligrams of volcanic ash for every cubic meter of air, an engine ingests 4.5 kilograms of ash every minute—which is enough to shut it down.

Volcanic ash is dangerous for hot turbine engines because it is very fine and melts easily. Flame-outs of jet engines occur because the melted silicate ash forms a thin coating of glass on the fuel injectors and on the metal surfaces of the turbine vanes. A second layer of ash fragments then adheres to the glass while it is still sticky. These layers, which may build up to several millimeters in thickness on the high-pressure turbine vanes, can cause the turbine to stall.

In addition to stalling the turbines, large volumes of ash sucked into the engines can abrade metal parts and surface coatings in the hot sections of the

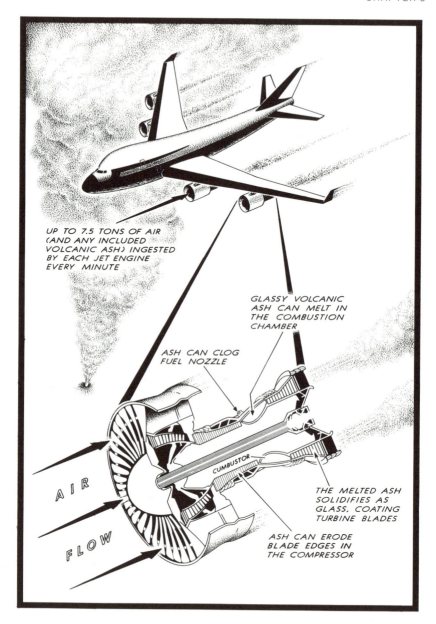

UP TO 7.5 TONS OF AIR
(AND ANY INCLUDED
VOLCANIC ASH) INGESTED
BY EACH JET ENGINE
EVERY MINUTE

GLASSY VOLCANIC
ASH CAN MELT IN
THE COMBUSTION
CHAMBER

ASH CAN CLOG
FUEL NOZZLE

THE MELTED ASH
SOLIDIFIES AS
GLASS, COATING
TURBINE BLADES

ASH CAN ERODE
BLADE EDGES IN
THE COMPRESSOR

AIR

FLOW

CUMBUSTOR

Fig. 8-4. What happens to the engines of a modern, high-performance jet, if it accidentally flies through a volcanic eruption plume? The engines of a Boeing 747 ingest
enormous volumes of air. If silicate and acid particles are dispersed in the air, they are
quickly concentrated by the engine compressor. The metal surfaces are quickly abraded

jet engines; contaminate plastic insulation and neoprene hosing in air distribution systems; corrode electrical contacts; and cause pitting, crazing, etching, and embrittlement of windows due to acid aerosols. The dielectric properties of some coatings that cover printed circuit cards may change because of the presence of the volcanic gases. Volcanic dust can damage landing lights, clog systems that help calculate airspeed and altitude for the pilots, and damage sensors that deliver electronic data to automated systems used to fly modern aircraft (fig. 8-4).

Ash and Aircraft Safety

In addition to causing jet flameout, volcanic ash is abrasive, mildly corrosive, and conductive when wet. It may carry a high static charge for as long as two days. Dry ash blown into the air reduces visibility and piles up on airport roads, runways, and taxiways. Fine ash penetrates all but the most tightly sealed enclosures and is very difficult to remove from electronic components. Wet ash is very heavy—up to 1,400 kilograms per cubic meter, and may cause buildings to collapse. Also, wet ash is very slippery and causes traction problems.

Ash must be physically removed and contained after removal to prevent reentrainment into the air by wind or nearby aircraft. Ash deposited upon electronic components, especially high-voltage circuits, can cause arcing, short circuits, and intermittent failure due to its conductivity. Ash dampened by rain can cause arcing, flashovers, and pole fires on electricity distribution systems. Electrical outages can, of course, have a severe impact on most airport facilities, including on runway lighting systems and on air-traffic control equipment.

In 1984, a committee made up of airline representatives, pilots, air-traffic controllers, a geologist, and a meteorologist was convened by the International Civil Aviation Organization to work on improving safety for aircraft operating downwind from erupting volcanoes and on educating pilots. The efforts of this committee and of hundreds of others in the aviation and volcanology communities since 1984 have changed safety regulations, have established education programs for pilots, have placed the prediction and tracking of eruption

---

and fuel nozzles become clogged. But most significantly, the operating temperatures within the engines (1,400°C) can melt volcanic glass particles. The melted ash then coats and sticks to the turbine blades, causing the engine to automatically shut down. The wrong thing for a pilot to do is gun the engines to escape, for it increases the temperatures. When the pilot realizes that the plane has entered an eruption cloud the right procedure, though counterintuitive, is to decrease power to the engines to lower the temperatures below the melting point of glass shards. (Adapted from Casadevall, 1992)

plumes on the same daily routine as weather forecasting, and have integrated knowledge of the hazard into the design of new aircraft. Other such efforts include a symposium, "Volcanic Ash and Aviation Safety," that was held in Seattle, Washington, in 1991, coincidentally one month after the Mount Pinatubo eruption, and a 1993 workshop on the impact of volcanic ashfalls on airports that was attended by volcanologists, flight controllers, and aviation meteorologists.

Most flight crews can now recognize the telltale signs of an encounter with an eruption plume, including St. Elmo's fire and the pungent odor of sulfur gases. Pilots have been instructed to throttle back jet engines to lower their operating temperatures below the melting point of ash and to make a 180-degree turn to fly back out of the plume. This counters the intuitive reflex of pilots, which is to increase thrust and climb out of the cloud as quickly as possible. Increasing the thrust increases engine temperature to above the melting point of ash and climbing upward may cause the aircraft to remain in an eruption plume, for some plumes rise as high as 15,500 meters, which is higher than commercial aircraft can operate.

Despite knowledge of preventative measures, there have been encounters between eruption plumes and aircraft because large explosive eruptions occur rapidly and communications can be imperfect. A map superimposing the world's active volcanoes, wind patterns, and major international flight routes shows that many of them overlap. Flights from Europe or North America to Japan, Korea, or Southeast Asia are downwind from the many volcanoes of the Alaska Peninsula, the Aleutian Islands, Kamchatka, the Kurile Islands, and Japan. There are hundreds of active and young dormant volcanoes along this route. Most are not monitored and many are located in uninhabited areas that have extremely foul weather. Some eruptions in the Aleutians often remain undetected until satellite images are analyzed (fig 8-5).

Most of the solid ejecta from volcanic eruptions falls from the atmosphere within a few tens of kilometers from the source, but perhaps as much as one-

---

Fig. 8-5. *Top*: The highest elevations reached by eruption clouds between 1975 and 1985. Also shown are the elevation of the tropopause (the altitude at which air temperatures no longer decrease and begin to rise) and the most common operating altitudes for commercial jet aircraft. (Adapted from McClelland et al., 1989). *Bottom*: A section of the polar flight route from Europe and North America to Japan and Southeast Asia passes downwind from chains of active volcanoes in the Aleutian Islands, on Kamchatka, and in the Kurile Islands (Russia), and Japan. To demonstrate the potential problem, the April 30, 1981, eruption plume from Alaid Volcano, in the Kurile Islands, is shown. There is now excellent communication between volcano observatories, meteorologists, and international flight controllers about any eruptions occurring near air routes such as this one. (Adapted from Fox, et al., 1989)

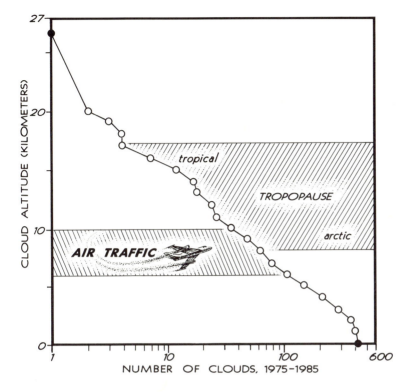

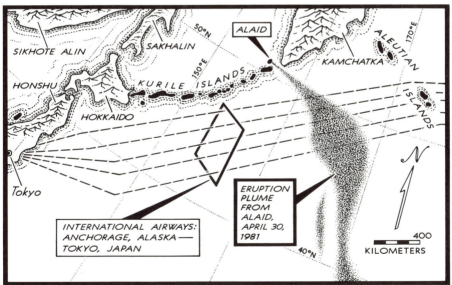

half of the ash can be fine enough to be carried far away from the eruption source by the wind. This latter occurrence is of major importance to the airline industry because widespread eruption plumes are dispersed downwind at jet-cruising elevations of 8–20 kilometers, and eruption plumes can drift for thousands of kilometers. An ash cloud from the eruption of Alaska's Mount Spurr on September 17, 1992, for example, drifted south and east across Canada and reentered U.S. airspace in the Great Lakes region, 4,000 kilometers from the source, and had major impact on air traffic. The base of the ash cloud was at 5,500 meters and its top as high as 13,600 meters. Pilots had to fly beneath the cloud rather than above or around it. Northern routes over the Great Lakes region therefore had to be changed to southern routes, causing disruption of airport schedules.

Although eruptions can be brief and in remote areas, potentially dangerous ash clouds can move rapidly from their source into populated areas and air corridors without being noticed until the ashfalls or aircraft are damaged; for example, a five-minute eruption of Lascar Volcano, in northern Chile, on September 16, 1986, produced a 15-kilometer-high eruption column that moved rapidly downwind as a 2-kilometer-thick eruption plume occupying elevations of 10–14 kilometers. The eruption plume passed over the city of Salta, Argentina, 285 kilometers from Lascar less than two hours after the eruption. Ash fell on the Salta airport, but there were no aircraft incidents. About three and a half hours after the eruption, the plume had traveled 400 kilometers and covered an area of 100,000 square kilometers.

It would be ideal to monitor all active volcanoes, but budget realities prevent such all-encompassing coverage, especially in remote areas. Perhaps the problem can be solved by modern technology because it is now possible to collect seismic and other data by battery-run or solar remote devices that relay messages through satellites to observatories. The information can be used to detect early signs of activity as well as for long-term monitoring. If there are signs of activity, observatories may then conduct more detailed observations and, if needed, advise the FAA or airline companies to modify air routes.

The coincidence of volcanoes and major transportation routes has resulted in more frequent encounters of people with explosive eruptions. And it is certain that the conflict between volcanoes and people will continue into the future, exacerbated in part by a steadily increasing world population.

WITHOUT the breath of volcanoes, there would be no air to breathe. The atmosphere has evolved mostly from exhalations of volcanoes, and the millions of tons of gases that pour into the sky each year from volcanic eruptions keep the atmosphere replenished. Without replenishment, gases of the atmosphere would gradually leak into space, molecule by molecule. But volcanic gases have more immediate effects as discussed in the next chapter.

References

Boeing Commercial Airplane Group. *Volcanic Ash Avoidance—Flight Crew Briefing.* Seattle: Boeing, Inc., 1993. Videotape 911202.

Casadevall, T. J. "Lessons of the past decade." *FAA Aviation Safety Journal* 2, no. 3 (1992): 3–11.

———. "The 1989–1990 eruption of Redoubt Volcano, Alaska: Impacts on aircraft operations." *Journal of Volcanology and Geothermal Research* 62 (1994): 301–16.

———, ed. *Volcanic Ash and Aviation Safety: Proceedings of the First International Symposium on Volcanic Ash and Aviation Safety.* U.S. Geological Survey Bulletin 2047. Washington, D.C. 1994.

Fox, T., G. Heiken, G. Sigvaldason, and R. Tilling. "Volcanic ash warnings for civil aviation." Abstracts for IAVCEI General Assembly. New Mexico Institute of Mining and Technology Bulletin 131. 1989, p. 96.

McClelland, L., T. Simkin, M. Summers, E. Nielsen, and T. Stein. *Global Volcanism 1975–1985.* Englewood Cliffs, N.J.: Prentice Hall, 1989.

Riehle, J. R., W. I. Rose, and D. J. Schneider. "Unmanned aerial sampling of a volcanic ash cloud." *EOS* 75 (1994): 137–38.

Simkin, T., and R. S. Fiske. *Krakatau 1883: The Volcanic Eruption and Its Effects.* Washington, D.C.: Smithsonian Institution Press, 1983.

Williams, H., and A. R. McBirney. *Volcanology.* San Francisco: Freeman, Cooper and Co., 1979.

The Breath of Volcanoes

Frontispiece. Volcanologists and their assistants are standing on the flat part of the active crater floor of Soufrière Volcano, Island of St. Vincent in the Lesser Antilles, in July 1979. The crater floor has been pushed upward from where the men stand and it slopes upward to the steaming vents at the base of the growing lava dome (dark gray area at top of photograph). The top of the lava dome (not visible in photograph) rises 100 meters above the crater floor. The seismologists are there to install posts at 100-meter intervals to visually monitor the growth of the dome from the rim of the volcano. The gases are mostly steam, and are venting to the atmosphere. (Photo: Richard V. Fisher)

Effects of Volcanic Gases

Except for abundant free oxygen (which is released via photosynthesis from plants, algae, and bacterium-like organisms known as cyanobacteria), all atmospheric gases come from inside the earth. Gases within magma remain dissolved because of the high pressures beneath the earth's surface, but as magma rises, the pressure is reduced and the dissolved gases expand and escape, like gases from well-shaken soda pop when the cap is removed. Water vapor constitutes 70–95 percent of all eruption gases. The rest consists of carbon dioxide, sulfur dioxide, and traces of nitrogen, hydrogen, carbon monoxide, sulfur, argon, chlorine, and fluorine.

The different volcanic gases affect the earth, its people, and its animals in various ways. Water vapor is generally beneficial, adding to the earth's water supply. Sulfur dioxide can form an aerosol that reflects the sun's heat rays and cools the earth's surface or causes harm by forming acid rains. Fluorine gases and their acid aerosols can be lethal to animals. Carbon dioxide can add to the effects of global warming and is lethal in high concentrations. And there is circumstantial evidence that volcanic eruptions can affect weather patterns and possibly trigger climatic change.*

Benjamin Franklin made a leap of intuition by noting a circumstantial link between the weather and the volcanic eruption of Laki Fissure, Iceland, in 1783. A blue haze extended over Iceland and northern Europe during that year, and Franklin suggested that the severe winter of 1783–1784 resulted from blocking of sun's rays by fine ash and gases. One hundred years later, following the 1883 eruption of Krakatau Volcano, a commission was appointed to study the affects of volcanism on the weather. Their results linked brilliant sunsets and other optical effects to the dust in the atmosphere from the eruption and described the global spread of the ash cloud. Unseasonably cool weather followed the eruption, but the evidence of cause and effect was circumstantial.

With the world's population expected to be twelve billion by 2035, the

---

* The words *climate* and *weather* are often used synonomously, but *climate* refers to the average weather conditions of a place as determined by the temperature and meteorological changes over a period of years. *Weather* refers to daily changes or weekly and monthly patterns; for example, we don't refer to a single rainstorm or a series of rainstorms as a bad climate. If it is excessive, it is bad weather.

effects of man's activities upon global climate become increasingly important to understand. We need to know what can alter the climate. Besides upsetting agricultural zones and planting seasons, for example, one of the main concerns about global warming is that it could cause ocean levels to rise if the glaciers in Antarctica melt. This would seriously affect many large, coastal population centers—London, Tokyo-Yokohama, Los Angeles, New York, Buenos Aires—as the water levels rose. If there is some evidence that man's pollutants, beyond that of natural contaminants such as volcanic gases, can cause global warming, steps must be taken to reverse the trend.

An analysis of the problem is exceedingly complex, for volcanoes can help cool the earth's surface by forming sulfuric acid aerosols that reflect and diffuse the sun's rays and also can contribute to global warming by pumping carbon dioxide into the atmosphere, exacerbating what is known as the greenhouse effect. The sun's rays enter the greenhouse through glass or plastic, but the heat produced cannot escape and is retained inside. Similarly, carbon dioxide behaves like a glass shield over the earth. The sun's rays can penetrate the carbon dioxide gas, but the carbon dioxide shield prevents heat from escaping into the atmosphere, causing temperatures to rise at the earth's surface.

There is scientific evidence that global temperatures have been steadily increasing since the beginning of the Industrial Revolution due to introduction of carbon dioxide to the atmosphere through the burning of fossil fuels by machines, industrial processes, and the internal combustion engine. Carbon dioxide is abundant in volcanic gases, but there is not enough to significantly contribute to the greenhouse effect—far more carbon dioxide is contributed by man's activities. Volcanoes contribute about 110 million tons of carbon dioxide per year whereas other sources contribute about 10 billion tons per year.

Small as it may seem, an average change of 2–5°C per year would greatly change the earth's climate. An average temperature decrease of this magnitude would eventually cause glaciers to grow or reform, thereby causing ocean levels to fall. An average temperature increase of this magnitude would eventually cause all glaciers on earth to melt and the ocean levels to rise. A small temperature change averaged over a year is hardly noticeable on a daily basis to an individual, and in fact could be masked by an excessively cold or hot winter followed by an excessively hot or cold summer, giving the impression that long-term global warming or cooling is not taking place.

Because volcanoes can contribute to either cooling or warming, they may add or subtract from whatever cooling or warming trends that currently exist from other causes, whether man-made or natural. For example, cooling can be caused by sunspots that reduce the amount of heat received from the sun at the same time that carbon dioxide from industrial activity is building up in the atmosphere and contributing to the greenhouse effect. The warming trend from the greenhouse effect would then be masked by the cooling effect of sunspots until the number of sunspots decreased. It is difficult to sort out the various factors because a cooling event caused by volcanic activity can temporarily counteract a man-made warming effect.

Irrespective of the causes of warming and cooling trends, we can take advantage of volcanic eruptions, treating them as atmospheric experiments. The time and volume of eruptions can be directly monitored, and the ash that pours into the atmosphere can be followed by ground sightings and remote sensing devices from satellites. The data can then be used to improve global climate models for the study of weather modifications that result from all causes, not just from volcanoes. The improved models can help evaluate and identify the impact of man's industrial wastes upon the environment—the greenhouse effect and destruction of the ozone layer by chlorofluorocarbons (CFCs) from aerosol sprays.

The greatest disturbance that volcanoes have upon the earth's weather patterns results from the sulfur dioxide they produce. In the cold lower atmosphere, the gas is converted to sulfuric acid by the sun's rays reacting with water vapor to form sulfuric acid aerosols. The aerosol remains in suspension long after solid ash particles have fallen to earth.

In 1901 it was proposed that volcanic dust intercepted and absorbed solar radiation so that the earth's surface cooled, thereby causing glaciation. It has been found, however, that fine ash particles from an eruption column fall out too quickly to significantly cool the atmosphere for the extended period of time needed for glaciation, no matter how large the eruption; for example, the dust cloud from the great 1815 Tambora eruption lasted less than two years, and its effects upon the environment, though harmful to people, were short-lived.

On the other hand, sulfur aerosols last for many years, and the amount of sulfur produced by several historic eruptions shows a good correlation with the average temperature decrease in subsequent years (fig. 9-1). The close correlation was first established after the 1963 eruption of Agung Volcano, in Indonesia, when it was found that sulfur dioxide reached the stratosphere and stayed as a sulfuric acid aerosol. Between about 15–25 kilometers above the earth, sulfur dioxide was converted to an aerosol layer of tiny sulfuric acid droplets (about 0.1 micrometer in diameter). The global, life-affecting consequences of the sulfuric acid aerosol layer are enormous for it absorbs infrared radiation, thus cooling the troposphere, and scatters solar radiation back to space, warming the stratosphere (fig. 9-2). This process in turn can profoundly affect the weather and may contribute to climatic changes. The sulfuric acid aerosol layer gradually dissipates, but it is renewed by each eruption that is rich in sulfur dioxide. That this occurs was confirmed by data collected after the eruptions of El Chichón, Mexico (1982), and Mount Pinatubo (1991), both of which produced plumes that contained high concentrations of sulfur compounds as had Agung Volcano.

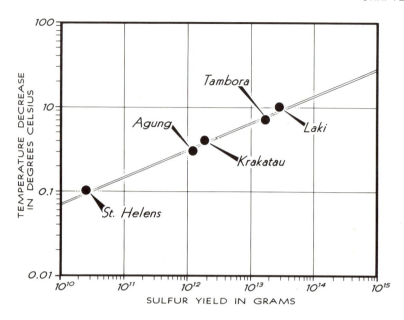

Fig. 9-1. Global temperatures decrease with an increase in sulfur emissions from volca-noes. Sulfur occurs as sulfur dioxide in eruption columns. Reduced temperatures can significantly affect the earth's weather patterns, as happened following the 1815 eruption of Tambora, Indonesia, and the 1783 eruption of Laki, Iceland. (Adapted from Sigurds-son, 1990)

The 1991 eruption of Mount Pinatubo, one of the largest of the twentieth century, is destined to become a watershed in the annals of meteorological research and of the effects of volcanic eruptions on climate because the sulfur-rich eruption coincided with modern environmental concerns and develop-ment of sophisticated instruments to monitor volcanic eruptions. Mount Pina-tubo's eruption produced the largest sulfur dioxide aerosol cloud of the last one hundred years. The data show that the stratosphere warmed up and the troposphere cooled down because of the replenishment of the sulfur dioxide aerosol.

Chlorine is another gas that can have negative effects upon the earth's envi-ronment. Satellite data suggest that ozone depletion can be exacerbated by volcanic chlorine, which is emitted as hydrochloric acid that then breaks down and forms chlorine and chlorine monoxide molecules. The sulfate aerosols furnish sites for the chemical reactions that release the chlorine atoms. This eruption-derived chlorine is added to the man-made chlorine already present in the stratosphere; all of the chlorine then proceeds to destroy ozone, each chlo-

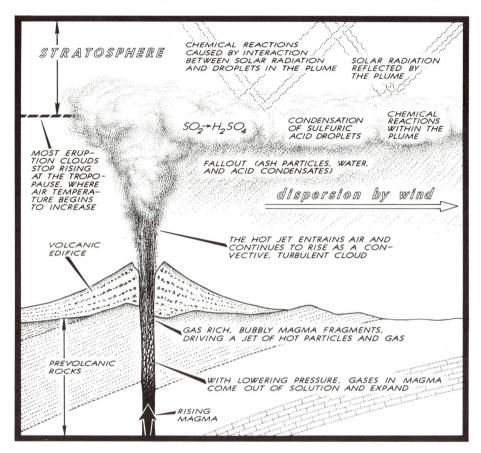

Fig. 9-2. The effect of explosive volcanic eruptions on the earth's atmosphere is a complex process involving interactions between gaseous, liquid, and solid volcanic products, the atmosphere, and solar radiation. (Adapted from an American Geophysical Special Report, *Volcanism and Climate Change*, 1992)

rine atom being recycled many times.* Satellite data from the Stratospheric Aerosol and Gas Experiment showed a 15–25 percent ozone loss at high latitudes after Mount Pinatubo's eruption and a reduction of nearly 50 percent over the Antarctic after the eruptions of Mount Pinatubo and Mount Hudson,

---

* The ozone layer, which begins at 12 kilometers above the ground in temperate latitudes, is a shield that protects living beings from the most harmful of the sun's ultraviolet (UV) radiation known as UV-B. In high enough doses, UV-B harms animals and plants by damaging cellular DNA. A well-developed ozone layer absorbs this radiation and protects organisms from these harmful effects.

Chile, in 1991. Ozone depletion by volcanic eruptions and use of CFCs by people combine to produce a great threat to ozone depletion, but we can control only man's contributions.

Another event may also obscure the impact of volcanic gases on the environment. El Niño, a periodic warm current along the west coast of South America, heats the earth's surface, but against a background of volcanism, its exact effects are difficult to sort out. El Niño is preceded by a sudden change in atmospheric and oceanic circulation in the equatorial Pacific, and the events, each of which lasts about a year, occur every three to seven years. Although El Niño is short-lived, it may trigger longer lasting changes in the weather by warming the earth's surface as the reflection of the sun's rays from volcano aerosols cool it; the effects of the two events, if they occur simultaneously, may mask one another. El Niño has followed numerous eruptions, so the specific impact of each type of event is unclear. Some scientists believe that volcanic stratospheric aerosols may trigger or intensify El Niño, whereas most workers believe they do not.

Another volcanic gas that can cause extensive problems is fluorine. It can condense in rain or on ash particles that fall on grass and into streams and lakes, polluting them with excess fluorine. Animals that eat grass coated with fluorine-tainted ash are poisoned. Small amounts of fluorine can be beneficial, but excess fluorine causes fluorosis, an affliction that eventually kills animals by destroying their bones. The Lakagigar, Iceland (Laki), eruption of 1783 produced fluorine-rich gases that eventually killed 79 percent of the sheep in Iceland, leading to the starvation and deaths of about ten thousand people, 20 percent of Iceland's population.

Some volcanoes deserve their notoriety, but others get mistakenly blamed as did Hudson Volcano in Chile in August 1991. Eruption plumes swept southeast across the ranch lands of Santa Cruz Province in southern Argentina, and an estimated 2 cubic kilometers of fine-grained volcanic ash fell onto grazing lands across Argentina. At 700 kilometers downwind of the volcano, so much ash fell and was blown about by wind, that visibility dropped to a few hundred meters. Fluorosis was initially blamed for killing large numbers of sheep in Santa Cruz Province during the following months, but a study by the U.S. Centers for Disease Control found that the deaths were provoked by a combination of harsh winter, overgrazed land, and complications of spring pregnancies. The volcano was not totally blameless, for all of these conditions were exacerbated by ash in the air and on grasses, but fluorosis was not the culprit.

The Year without a Summer

The eruption of Mount Tambora in April 1815, believed to be the largest of the last ten thousand years, is an example of how difficult it is to identify the causes of climatic changes, for this eruption occurred during a time of visible sunspot activity and during a cold period that had started in 1811 and ended in 1817. The eruption exacerbated the global cooling that had been initiated as a result of another cause. The atmospheric effects of Mount Tambora's eruption also demonstrate that the earth is a holistic system—the cold period caused famine and economic disruption in the northern hemisphere, thousands of miles from Tambora Volcano, which is in the southern hemisphere.

Mount Tambora's eruption took place on the island of Sumbawa, Indonesia. The noise was heard 2,000 kilometers from the volcano, and the ash that fell on Java, 483 kilometers away, was several centimeters thick. About 12,000 inhabitants of Sumbgasa perished as a direct result of the eruption and 44,000 people died on the neighboring island of Lombok, primarily from famine resulting from the destruction of crops by falling ash. Circumstantial evidence suggests that the enormous aerosol volume sent into the atmosphere by the eruption caused anomalous summer weather the following year. In the northern hemisphere, parts of western North America, eastern Europe, and Japan had average or above-average temperatures, but there was remarkable cold over much of eastern North America, western Europe, and China. The southward flow of freezing Arctic air in one region was offset by poleward flow of tropical air in another.

In addition to the recorded events of the time, tree rings indicate that reduced temperatures followed the eruption, and the acidity of ice layers in the Greenland and Antarctica ice caps are evidence for abundant sulfur in an upper troposphere aerosol layer. In eastern Hudson Bay, Canada, the mean daily temperature in midsummer 1816 was reduced by about 5 to 6°C. In New England, on June 6–11, July 9, August 21, and August 30, 1816, there were snowfalls and killing frosts that destroyed all but the hardiest grains and vegetables. The greatest loss stemmed from the failure of the corn crop, mainly because it was used as fodder for cattle, hogs, and chickens. The failure of Canada's wheat crop resulted in a shortage of bread and milk. In New York, soup kitchens were opened to feed the starving. The cost of wheat, grains, and flour rose sharply in New England, Canada, and western Europe. At the same time the cost of beef and pork went down as cattle and pigs were slaughtered because farmers couldn't afford to feed them. Sea ice obstructed ships in the Atlantic shipping lanes, and glaciers advanced in the mountainous regions.

A medical account printed in 1817 by Dr. Thomas D. Mitchell blamed the thermal deficit on volcanic dust in the atmosphere:

> What rendered it more astonishing in its diurnal variation, was its coexistence with mist or vapour equally dense and diaphanous all over the horizon. It had nothing of the nature of a humid fog. It was like that smoking vapour which overspread Europe about thirty years ago [during the Laki Fissure, Iceland eruption in 1783]. The learned, who made experiments to ascertain its nature, could only state its

remarkable dryness. . . . While the human eye could thus, during the long days, gaze on the great luminary of nature [the sun], . . . deprived of its dazzling splendour and radiance; then the numerous dark spots were discovered on its face, without the help of telescopic or obscured glasses; these among the multitude became the theme of popular apprehension of a calamitous sign in heaven, and others thought to have found a visible cause of the long refrigeration of our atmosphere. (quoted in Post, *Subsistence Crisis*, p. 25)

Other atmospheric phenomena occurred because of volcanic dust in the atmosphere, including a red, blue, or green sun and moon, richly colored twilights, dimming of the sun at the horizon, and colored snow. There was a great Hungarian blizzard in 1816 that produced brown or flesh-colored snow, the people of Taranto in southern Italy were alarmed by red and yellow snow, and snowfalls in April and May 1816 in Maryland were tinted brown, bluish, and red.

Europe had an even more disastrous summer than North America. July 1816 was the coldest July in 192 years in the English Lancashire plain, and the mean summer temperature in 1816 at Geneva, Switzerland, was the lowest since 1753, ruining that season's harvests and leading to famine throughout the country. Summer wheat could not be replanted because there was no seed in Switzerland's granaries, and pigs had to be slaughtered because there was no fodder. People were forced to eat potatoes, a food source looked down upon as food for peasants, but nevertheless, by August the potatoes were gone. By mid-1817, the price of grain in Switzerland had tripled, Zurich was overrun by beggars, and parish records registered many deaths from starvation. In Sorrel, Iceland, moss and cats were eaten.

Partly from the political ferment stemming from Napoleon's defeat at Waterloo in 1815 and partly as an outgrowth of the cold summer, poor harvest, and subsequently high food prices in 1816, there were riots in France. Rioting and crime increased in the winter of 1816–1817 as prices continued to rise and grain supplies dwindled. In villages across France, the people battled authorities for the meager grain supplies. Farmers carrying wheat to market in villages along the Loire Valley required protection by soldiers and police. So many food-related crimes were being committed each day by December 1816, that overwhelmed authorities had to practically ignore them.

Mount Tambora's eruption had a strong domino effect. The difficulties of the European governments with their waves of social and political unrest were the end point of a connected sequence of events that began with the disruption of weather patterns by volcanic dust and aerosols in 1815. Anomalous weather patterns of 1816 caused a sharp decrease in agricultural productivity, especially of grains, with the resulting scarcity and high price of food. These effects set in motion inflation, unemployment, famine, disease, political unrest, and riots.

The causes of climatic changes, including the onset of the ice ages, are still unknown. Certainly, climatic changes so long ago were not man-made but resulted from natural causes, a prime suspect being volcanism. There may be triggering mechanisms that can significantly shift global temperatures, but the likelihood of dramatic change is increased by man's atmospheric pollutants. Which has the greater potential to affect future climatic change, man or nature? This critical question cannot be completely answered until the relative contributions of each is known. But by the time it is scientifically proved, one way or the other, it may be too late to affect the outcome.

NOT all future global atmospheric changes will cause physical harm to people, but they may cause enormous economic loss; for example, if the sea level rises a few feet, many major cities on earth will have to be moved or enormous dikes built to protect them. But whatever changes may occur, humans, animals, and plants will have to adjust to new environmental conditions. In part 3, we will focus upon aspects of volcanoes that are beneficial, and at times, downright profitable.

References

Appenzeller, T. "Ancient climate coolings are on thin ice." *Science* 262 (1993): 1818–19.

American Geophysical Union. *Volcanism and Climate Change.* AGU Special Report. Washington, D.C.: American Geophysical Union, 1992.

Harington, C. R., ed. *The Year without a Summer?* Ottawa: Canadian Museum of Nature, 1992.

Post, J. D. *The Last Great Subsistence Crisis in the Western World.* Baltimore: Johns Hopkins University Press, 1977.

Sigurdsson, H. "Assessment of the atmospheric impact of volcanic eruptions." In *Global Catastrophes in Earth History; An Interdisciplinary Conference on Impacts, Volcanism, and Mass Mortality,* edited by V. L. Sharpton and P. D. Ward, 99–110. Geological Society of America Special Paper 247. Boulder, Colo., 1990.

Stommel, H., and E. Stommel. *Volcano Weather: The Story of 1816, the Year without a Summer.* Newport: Seven Seas Press, 1983.

Williams, H., and A. R. McBirney. *Volcanology.* San Francisco: Freeman, Cooper and Co., 1979.

WOVO [*World Organization of Volcano Observatories*] *News*: no. 1 (winter 1993): 1–15.

Myths and Benefits
of Volcanoes

CHAPTER 10

Sometimes the Gods
Are Angry

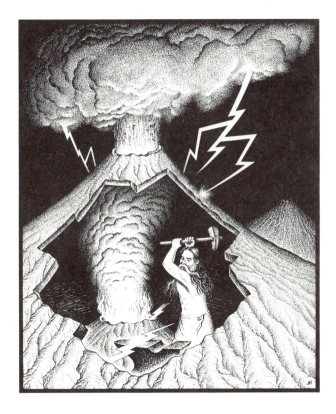

Frontispiece. Vulcan worked at his forge within a volcano on the tiny island of Vulcano in the Aeolian Islands, Italy. His nasty temperament and fits of rage
were greatly feared. He forged weapons of war for
the gods, and without provocation, Vulcan could destroy people and their property with fire and lightning, lava flows, or volcanic explosions.

Fig. 10-1. Crater of Vulcano, on Vulcano Island, Aeolian Islands, Italy. The Romans believed that Vulcan, the god of the underworld, lived below this intermittently active volcano. "Vulcan" is the root for the word *volcano*. (Photo: Richard V. Fisher)

The Roman god Vulcan was the unruly child of Jupiter and Juno; he lived with his forge inside a volcano on the tiny island of Vulcano in the Aeolian Islands, Italy (fig. 10-1). Volcanic eruptions were the ash and glow produced from his forge as he fashioned bronze arrows for Apollo, armor for Hercules, and thunderbolts and lightning for Jupiter. Without provocation, Vulcan would terrify and even destroy people and their property with fire and lightning, lava flows, or volcanic explosions. It was from the island of Vulcano that the word *volcano* was incorporated into the English vocabulary. In ancient Rome, Vulcanalia was a celebration held each year on August 23 with rituals and sacrifices to Vulcan to prevent disasters from fire, lightning, and volcanic eruptions. Sometimes it helped, and sometimes it didn't, but just in case it would, it was best not to forget that special day to forestall his anger.

The ancient Greeks thought that Zeus buried giants beneath the mountains and that volcanic eruptions were caused by their heavy breathing as they tried to escape. Virgil wrote that the eruptions of Mount Etna were the burning breath of the Titan Enceladus as he struggled to free himself from the underground prison to which he had been sent by Jupiter. The earthquakes were from Titan's exertions and the rumblings were his angry voice.

Violent shaking of the earth from a powerful earthquake, gigantic waves in a stormy sea, and the roaring power of a volcanic explosion were unexplainable to ancient people, who offered sacrifices and prayers to assuage the powers of the gods who were thought to cause such catastrophic events. The people invented stories, composed songs and poems dedicated to the gods, and practiced rituals to soothe their anger.

Volcanoes Assault the Senses

Volcanoes assault all of the senses—sight, touch, smell, and hearing. One can even taste the sulfur dioxide that wafts through the air. But it is the sound, combined with a shaking earth, that causes stark terror. The relentless roaring of a volcano mimics that of a NASA rocket taking off or the unbearable noise close to a jumbo jet at full throttle. Through the night, sounds can continue unabated for hours and spook the darkness with unknown explosions, pops, whistling, and swishing sounds like ghostly winds through a pine forest. The roar of the phreatic (steam) eruptions of Soufrière de Guadeloupe in 1976 caused some of the volcanologists on the volcano to raise their eyes skyward to see if there were low-flying jet aircraft.

In 1883 on Rodrigues Island, which is east of Madagascar in the Indian Ocean, the people heard noises resembling artillery fire from the British navy, only to find out later that they were explosions from the eruption of Krakatau, Indonesia, some 4,800 kilometers farther east. The eruption gained notoriety as the loudest noise ever to have occurred on earth.

British and Dutch scientists were intrigued by the sounds generated during the eruption of Krakatau because of their wide distribution. Exceptionally loud noises were heard over the entire Indonesian archipelago as well as in areas much farther away. But close to the eruption there were quiet zones, which were believed to have been caused by sounds that were poorly transmitted through dense ash clouds. Inaudible shocks were sensed by barometers and tide gauges, and they reportedly caused temporary deafness or buzzing in the ears. Shock waves circled the earth several times, measured by abrupt changes in barometric pressure.

The somewhat smaller eruptions of Mount Pelée sound like large rocket engines, the noise being generated by the turbulence of escaping gases. Occasionally an explosion within the upper part of a vent will generate a shock wave called a *flashing arc*, which is caused by gases escaping at supersonic velocities. During one explosive event in the eruption of Mount Pelée in 1930, Frank Perret, a pioneering American volcanologist, had a close encounter with volcanology's most dreaded phenomenon, a pyroclastic flow. Perret had built an observation shack close to the valley through which pyroclastic flows had moved toward the sea from the volcano. One barely missed his shack as he watched, and so gave us the first eyewitness report by an experienced volcanologist about the sound from a pyroclastic flow. He reported that there was no sound! Perret surmised that expanding gases from individual fragments kept them from colliding and thereby stifled the noise. In 1906, Frank Perret had pioneered the study of volcano sounds by recording the eruption of Mount Vesuvius, Italy, using a microphone and a wax recording instrument.

Ray Wilcox, an American geologist with the USGS who was assigned as "permanent" observer to the eruption of Parícutin Volcano, Mexico, during the 1940 eruption, was the first person to distinguish the sounds of a variety of volcanic processes. Surging or swishing noises, barely audible on the volcano slopes, accompanied the ash and pumice particles falling through the forest. A sound like that of beating surf accompanied successive bomb bursts from the crater, and then following that were the slapping sounds of bombs falling onto the volcano's flanks. Gases from small vents produced a low growl when they escaped intermittently and a deep throaty roar with increased flow rates. Intensely loud vapor columns from the crater produced a noise akin to rolling thunder. When lava was being emitted, it was accompanied by rhythmic sounds like those of a steam locomotive.

Pacaya Volcano, near Guatemala City, has had intermittent Strombolian eruptions for decades. When molten bombs are ejected, there is a raucous clatter, like someone hammering on a corrugated steel roof. In contrast, when bombs, ash, and dense steam are erupted together, there is little more than a subdued "whoosh." Recent studies demonstrate that the Strombolian clatter is generated as large gas bubbles break at the surface of the lava column.

An unforgettable musical sound on a quiet day in Hawaii near newly emplaced lava flows is the delicate tinkling of millions of glass shards, broken from the surface of the cooling lava and set into motion by the lightest of breezes.

Changing Ideas from Old
to New

Scientific inquiry in the Western world, started by the Greeks, returned in the fifteenth century A.D. following the Dark Ages in Europe and gradually led to new disciplines through which to understand nature, one of them being the science of the earth. A deceptively simple but revolutionary concept in the geological sciences was that the earth's surface changes at an extremely slow pace—on the order of a few millimeters or centimeters every one hundred years or so—and that vast geological changes are progressing today at about the same rate as they did in the past, yet they can raise mountains to great heights and carve canyons into mile-deep gorges, such as the Grand Canyon, over the short geologic span of a million years. Sudden geologic catastrophes, such as volcanic eruptions, tsunamis, earthquakes, and floods, punctuate the slow, background processes. Long-term geologic changes that occur by small increments have been formulated into the principle of Uniformitarianism; its underlying philosophy allows the interpretation of ancient rocks as explained below.

The life span of human beings is too short for us to perceive the slower changes that are taking place, although we can observe some of the means by which these changes occur. We can watch flooded rivers carrying loads of sediment and boulders that scrape and erode river bottoms, wind carrying dry sand relentlessly sandblasting whatever is in its path, waves on the beach building dunes, and volcanoes emitting streams of lava or pyroclastic flows. All of these processes form layers of sand and gravel, or lava and tuff. Comparisons of their deposits with much older, well-hardened layers seen on eroded edges of cliffs enable us to explain their origin. Thus, what we observe in action today provides the intellectual tools to interpret ancient rock layers. This process can be expressed as, "The present is the key to the past." It is also possible to apply the principle of uniformitarianism in reverse by analyzing geological features or rock layers that have no observational precedent in order to predict what may occur in the future. One example comes from the existence of large calderas (see chap. 5); we have never seen one form, but we can confidently declare that gigantic eruptions will create another large caldera in the future.

Myths and legends about the earth are commonly thought to be fiction, but grains of truth have been discovered in them. Most of the world's legends about the earth have a geological or meteorological event as their basis. It is said by the Klamath Indians that Llao, chief of the Below World, who was located on a volcanic complex known as Mount Mazama (the site of Crater Lake, Oregon), and Skell, chief of the Above World, who was located on Mount Shasta, were locked in a mighty battle. While darkness covered the land, they threw gigantic stones and fire back and forth at each other. The battle went on for years, but finally ended when the Mount Mazama complex of volcanoes collapsed under Llao and he fell back to his underground world, leaving a huge hole that filled with rain to form Crater Lake. The eruption leading to the formation of Crater Lake occurred about 6,900 years ago and was doubtless witnessed by many Native Americans of the region. The myth preempted the discovery by Howell Williams, an eminent volcanologist in the mid-twentieth century, that the area around what is now Crater Lake collapsed to form a huge crater, 8 kilometers in diameter. Before Williams's research, it was generally thought that Crater Lake was a volcano that had its top blown away in a gigantic explosion, yet it was the Indian legend that contained the seeds of truth.

Thera (Santorini): Its Influence on the Modern World

A prime case of a catastrophic geologic event linked to myth and to the human condition is the Bronze Age eruption of Thera, Greece. During this period, the Aegean was dominated politically and economically by seafaring peoples of the Minoan city-states on Crete and on the Greek mainland. But in the seventeenth century B.C., the Minoan influence rapidly declined, although the exact reason remains unclear. Some people believe that it stemmed from a sociopsychological demoralization that started with a catastrophic eruption on the island of Thera.

If the eruption, either directly or indirectly, caused the decline of the Minoans, it can be considered to be one of the most influential volcanic eruptions ever to have taken place in human history. Its great impact resulted from the fact that it was a large event in the central part of the relatively small (at that time) developing western world.

Thera, which is about 70 kilometers north of Crete, is one in a chain of volcanoes that parallels the boundary of the Aegean tectonic plate, where Africa is pushed beneath southern Europe. Before the eruption, Thera was an island that consisted of a volcanic field built upon the western flanks of a limestone mountain, the present-day Mount Profitis Ilias. Overlapping lava shields made up the northern half of the island and a large flooded crater dominated the southern half. This small but complex volcanic field had been growing sporadically for more than a million years. The aridity of Thera resulted in soils adequate only for growing beans, olives, and grapes. The strategic location of the island—between Crete and the mainland—may have fostered the growth of the sophisticated town of Akrotiri, which, along with the large harbor in a flooded crater, served as a haven from the sudden storms that occasionally swept across the southern Aegean Sea.

In 1939, the Greek archaeologist Spyridon Marinatos proposed that Minoan civilization ended with an enormous eruption, and thirty years later he began excavating Akrotiri, a well-preserved Bronze Age town on the southern tip of Thera that had been buried by pumice fallout and pyroclastic flow deposits of the eruption (fig. 10-2). Sometimes called the Pompeii of the Aegean, Akrotiri gives a remarkable snapshot of life 3,600 years ago, when the Minoan city-states were at the height of their power and influence. But suddenly, sometime in the seventeenth century B.C., Akrotiri and other settlements were either buried or swept away by products of Thera's eruption. The catastrophic eruption produced a caldera 6 kilometers in diameter that broke the northern half of the island into three islands bordered by a bay 300 meters deep. It is also entirely possible that the caldera and flooding of the crater by the sea caused tsunamis that swept the eastern Mediterranean shore, but direct evidence is lacking and it is considered a possibility only because an island volcano of similar size and a postulated similar magnitude (the 1883 eruption of Krakatau) generated large tsunamis that devastated the coastline of the Sunda Straits. Classical scholars have proposed that the sources for many flooding legends, especially those in classical Greek mythology, may have had their origins in tsunamis produced by Thera's eruption during the Bronze Age.

Originally it was thought that the eruption directly caused Crete's downfall by destroying the island through heavy ashfall and tsunamis, but that was not the case. The destruction of Crete's palaces came two hundred years later in the middle of the fifteenth century B.C. Also, there is very little ashfall on Crete and little substantial evidence of flood damage of the island's ports by tsunamis (fig. 10-3).

Even if this particular eruption of Thera did not end Minoan civilization, it was a catastrophic event that has reverberated through time. For, as was proposed by the archaeologists Henry Schliemann in the late 1800s and Martin Nilsson in the 1930s, natural and historic events such as Thera's eruption, provided the Bronze Age peoples with the incidents upon which Greek mythology is based. Some scholars have proposed that the Bronze Age eruption of Thera is the basis for Plato's legend of the lost continent of Atlantis—a large continent that supposedly sank beneath the Atlantic Ocean—which is still believed by some people to this day.

Fig. 10-2. *Top*: Akrotiri, a Bronze Age town, on the island of Thera (Santorini), Greece, was buried by a large eruption in the seventeenth century B.C. Multiple-story buildings, colorful frescos, and other items were preserved by the pumice fallout from this eruption, giving Akrotiri the title Pompeii of the Aegean. In this view, storage jars lean against the outer wall of a townhouse of the time. (Photo: Grant Heiken); *Bottom*: One of the more well-preserved houses within Akrotiri, which was covered by thick pumice fallout before it was excavated. The top story of the building was toppled by volcanic surges originating at the eruption center, which was north of Akrotiri. The concrete pillars are modern, emplaced to stabilize the excavated house in this seismically active part of the Aegean. (Photo: Grant Heiken)

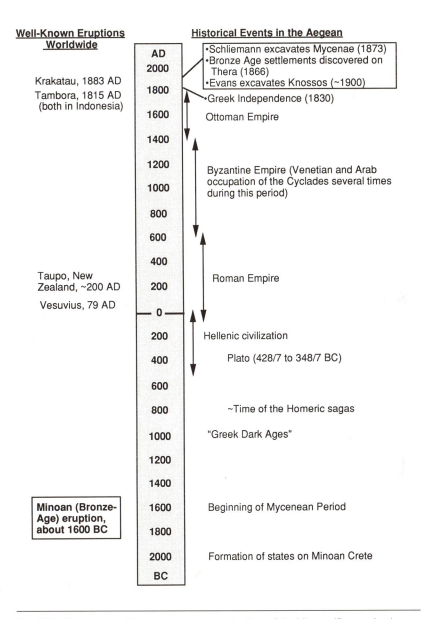

Well-Known Eruptions Worldwide		Historical Events in the Aegean
	AD **2000**	•Schliemann excavates Mycenae (1873) •Bronze Age settlements discovered on Thera (1866) •Evans excavates Knossos (~1900)
Krakatau, 1883 AD	**1800**	•Greek Independence (1830)
Tambora, 1815 AD (both in Indonesia)	**1600**	Ottoman Empire
	1400	
	1200	Byzantine Empire (Venetian and Arab occupation of the Cyclades several times during this period)
	1000	
	800	
	600	
	400	
Taupo, New Zealand, ~200 AD	**200**	Roman Empire
Vesuvius, 79 AD	**0**	
	200	Hellenic civilization
	400	Plato (428/7 to 348/7 BC)
	600	
	800	~Time of the Homeric sagas
	1000	"Greek Dark Ages"
	1200	
	1400	
Minoan (Bronze-Age) eruption, about 1600 BC	**1600**	Beginning of Mycenean Period
	1800	
	2000	Formation of states on Minoan Crete
	BC	

Fig. 10-3. Chronology of the Aegean, showing the time of the Minoan (Bronze Age) eruption in the seventeenth century B.C. and other eruptions and important historic events. Nearly eight hundred years elapsed between this major natural calamity and when the Homeric sagas were comprised. The eruption was about twelve hundred years before the beginning of Hellenic civilization. It has been proposed that many tales of the Hellenic gods are based on the disastrous effects of the Bronze Age eruption, with the stories passed on and embellished through time by oral histories and traditions.

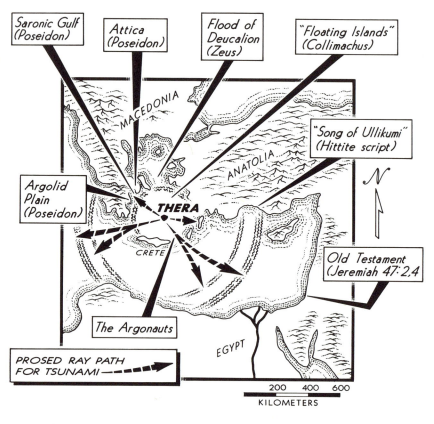

Fig. 10-4. Proposed movement of tsunamis generated by the Bronze Age eruption of Thera and the locations where legends concerning a great flood developed. Such legends are common in tales of the Greek gods, in the Old Testament, and in Hittite stories.

The flooding legends of the eastern Mediterranean may have had their origins along the coastline of Attica (where Athens is located), the Argolid Plain (in the Peloponnese, near Mycenae), the Saronic Gulf (immediately west of Athens), and the Anatolian (Turkish) coast (fig 10-4). These legends have been summarized by J. V. Luce, a well-known classics scholar at Trinity College, Dublin, in his book *The End of Atlantis*, and by the geophysicist James Mavor in his book *Voyage to Atlantis*. The best-known of these legends is that of Deucalion's flood. In this story, Zeus was angry with the rowdy, irreverent sons of Lycaon, so, disguised as a poor traveler, he paid them a visit, and they served him a stew of goat and sheep entrails. Enraged, he returned to Mount

Olympus and proceeded to flood the world with the intention of drowning all mankind. The titan Prometheus warned his son, Deucalion, of the impending disaster. Deucalion and his wife, Pyrrha, built a boat to ride out the flooding and thus saved the species, a story that presages the biblical story of Noah and the ark.

Other fits of pique among the Greek gods were caused by land disputes between Poseidon and Athena. Poseidon was a bad loser, so he flooded the lands around the Saronic Gulf and made them unfruitful for a time. In another legend based in Attica, Athena created an olive tree and Poseidon created a spring. Athena's invention was judged to be the more useful—so Poseidon again flooded the countryside. The flooding legends were passed from generation to generation by word of mouth until they were recorded in writing. Some of the legends were probably rooted in catastrophes from ordinary floods, but perhaps at least one had its origin in a long-forgotten volcanic eruption that caused tsunamis.

The return of Jason and the Argonauts to Greece with the golden fleece also hints of an ancient eruption in the region. The Argonauts were assaulted by Talos, a bronze giant, who threw fragments of stone at them. A modern volcanologist, Roberto Scandone of the University of Rome, has another explanation. Sailing past Rhodes and Karpathos to the east coast of Crete, the Argonauts were directly under the pumice and ash fallout from the Bronze Age eruption of Thera (the pall of darkness). They spent a night on the northeastern coast of Crete and then sailed northward downwind from Thera during the fallout of the last eruption phases (Talos throwing stones). Beyond the fallout, they anchored at the island of Anafi, gave thanks to Apollo for their survival, and then sailed home.

Volcanoes and the Underworld

Volcanic eruptions are terrifying events, and to a prescientific world, it was reasonable to ascribe them to subterranean supernatural forces. Craters and fissures were considered entrances to the underworld by all cultures— from Iceland to New Zealand and from West Africa to Indonesia. The underworld was commonly inhabited by gods of fire and mythical blacksmiths, but not all of them were forces of evil or retribution. The Greeks, for example, considered that losses during eruptions were unpredictable and inevitable— an attitude that permeates most cultures even today—but to help circumvent the losses, they climbed Mount Etna and offered incense to appease the gods.

To the early Christians, eruptions were punishment for a sinful life. The fire and brimstone of a Mount Etna eruption is the embodiment of a Christian hell conveyed for several hundred years from countless pulpits.

Fig. 10-5. Mount Fuji is the symbol of Japan and has been worshipped by millions of Japanese for many centuries. This photograph is from a lantern slide taken in the late nineteenth or early twentieth century. (Photo: source unknown)

The Japanese attitude toward volcanoes is cordial, or at least understanding (fig. 10-5). According to Daniel Boorstin,

> Their reverent, friendly, and intimate Japanese feelings toward nature have been expressed in an attitude to mountains very different from ours in the West. Japanese country folk have viewed mountain peaks and even smoking volcanoes as collaborating spirits.

> Cults of the mountain kami (spirits) as early as the Nara period (710–794) nourished the supernatural powers by mountain asceticism and bred belief in mountain magic. The flourishing cult of Mount Fuji (now with more than thirteen hundred shrines) made its majestic volcanic cone the nation's symbol.

> . . . There is no period of recorded history when the Japanese were not climbing Mount Fuji. Its symmetrical cone was one of the oldest subjects of their art and

poetry. The ascent of Mount Fuji with its ten stations early became a ritual, and the circuit of the crater's rocky peaks carried a high ceremonial meaning of Japanese affinity with nature. (Boorstin, *The Creators*, pp. 136–37)

Divine Intervention

In a cartoon by the artist Gary Larson, a group of generic natives are carrying two middle-aged female tourists toward a smoking crater. The caption is "Ha! And you were worried they wouldn't like Americans—Why, these people just lit up when I explained we were Virginians." The sacrifice of virgins to appease the volcano gods is a common perception that cartoonists and screenwriters have of people in ancient times living near volcanoes. In the Hollywood movie *Joe and the Volcano*, the hero is hired for sacrifice in exchange for a pleasurable, if short, life.

There have been cultures in which human sacrifices were made to quell the fury of gods or goddesses who resided within volcanoes. These rites have been practiced at volcanoes from South and Central America to Java and Africa, and the sacrifices have ranged from babies to numerous male slaves. Sacrifices are still made today, but not of people, and can include a few head of cattle or a bottle of brandy.

In Hawaii, gifts such as flowers, candy, fruit, cloth, and alcoholic beverages (usually gin) are offered. Older Hawaiians telling stories of Madame Pele recount the history of the Hawaiian Islands. They say that the islands are made of volcanoes that become younger from west to east, a story that predates confirmation by modern radioactive dating methods. Pele carried a magic digging stick that opened up volcanic craters wherever she dug. In chronological order, she first created the volcanoes of Kauai, then of Oahu, Molokai, Maui, and then of Hawaii, where she dug her last fire pit, Halemaumau, on the floor of Kilauea Caldera. She and her numerous relatives went to live in Halemaumau, which is described in the verbal tradition of Hawaii and recounted by a British missionary named William Ellis: "Halemaumau . . . had been burning from time immemorial . . . and had overflowed some part of the country during the reign of every king that had governed Hawaii; that in earlier ages it used to boil up, overflow its banks, and inundate the adjacent country; but that for many kings' reigns past it had kept below the level of the surrounding plain, continually extending its surface and increasing its depth, and occasionally throwing up, with violent explosions, huge rocks, or red-hot stones" (quoted in Vitaliano, *Legends*, p. 109). Madame Pele is very temperamental and easily given to anger, like the Roman god Vulcan. When she is displeased, she stamps her foot to start an eruption, which also causes earthquakes, and sends out rivers of lava to destroy the source of her displeasure, often killing many innocent victims.

In August 1881, when lava from Mauna Loa was threatening Hilo, Princess Ruth Keelikoani, granddaughter of King Kamehameha, was asked to arbitrate with Madame Pele. The 400-pound princess was carried to a place near the edge of the advancing lava flow, and as she chanted ancient incantations, she scattered red silk scarves on the lava flows and sprinkled a bottle of brandy onto the lava and burning grass. It worked! The lava stopped short of the city on the following day.

Madame Pele lives on in the twentieth century, and without doubt, will survive well into the twenty-first. In 1955, as a lava flow threatened the village of Kapoho, villagers chanted beside the lava stream and offered food and tobacco to her. The lava flow stopped short of the village that time. But Madame Pele returned in January 1960, and nothing the villagers could offer would stop her (fig. 10-6).

In medieval times, the Christian Church proposed that hell was below the earth's surface and that volcanoes were gateways to the nether world. Eruptions were the fires escaping from hell and the sounds were voices of the damned. The sulfurous stink added a touch of the macabre. One of the more popular entrances to hell was the crater of Mount Hekla in Iceland, a perception strengthened by St. Brendan, who may have seen an eruption there during the sixth century A.D. In his monograph on Mount Hekla, Sigurdur Thorarinsson, a distinguished Icelandic volcanologist, cites the poem *Voluspa* (The Sybil's Prophecy), which was composed in Iceland about the time when the country was converted to Christianity in A.D. 1000. One stanza suggests that the poet had witnessed a volcanic eruption, possibly at Katla.

The sun grows dark
the earth sinks into the sea,
the clear bright stars
disappear from the sky;
vapour pours out
and fire, life's nourisher:
the high flame
plays on heaven itself.

Fig. 10-6. *Top*: Kapoho Village, Hawaii, was a small village with a general grocery store before its destruction in 1960 by lavas from the east rift of Kilauea. Eruptions started from near sea level along the rift. This photograph shows steam plumes (white) from interaction of water with the magma side by side with magmatic lava fountains (gray). (Photo: *Honolulu Advertiser*). *Bottom*: A small cinder cone was built in back of Kapoho Village, but the town was destroyed by lava flows and ash burial. This photograph was taken near the entrance of the houses shown in the photo above. (Photo: *Honolulu Advertiser*)

Fig. 10-7. Tile painting of San Gennaro, patron saint of Naples. Here, he is saving the city from one of the Vesuvius eruptions. It is believed that the relicts of San Gennaro (dried blood and bones) have the power to protect Naples from eruptions of Mount Vesuvius. (Photo of tile by Richard V. Fisher)

Thorarinsson thinks that "it is undeniable that a great Katla eruption and the adoption of Christianity offered in combination no small cause for meditation on the day of doom" (Thorarinsson *Hekla*, pp. 21–22).

The link between Christianity and the belief that faith can stop a volcanic eruption is well illustrated at Mounts Vesuvius and Etna, where medieval traditions developed and flourished through time. There is no better modern example of volcano hazard mitigation by faith than in Naples. In the Gothic Duomo of San Gennaro is a richly decorated chapel that contains the skull of St. Januarius (San Gennaro); also in the tabernacle are two vials of his blood. St. Januarius, who may have been a native of Naples, was Bishop of Benevento when Diocletian, the Roman Emperor, from A.D. 285 to 305, began his persecutions (fig. 10-7). St. Januarius and two companions were to be eaten by beasts in the amphitheater, but none of the animals would touch them. The martyrs were then beheaded. St. Januarius was later beatified and his remains were kept in Naples. His sainthood had nothing to do with Vesuvius, but when the volcano erupted, the local people appealed to their patron saint to stop it.

In the fifth and sixth centuries, eruptions of Mount Vesuvius threatened Naples with ash clouds:

> The intercession of St. Januarius was implored at Naples on these occasions, and divine mercy so wonderfully interposed in causing these dreadful evils to cease thereupon, especially in 685 [explosive?] eruption. Bennet II being pope, and Justinian the Younger emperor, that the Greeks instituted a feast in honor of St. Januarius with two yearly solemn processions to return thanks to God. The protection of the city of Naples from this dreadful volcano by the same means was most remarkable in the years 1631 [highly explosive; pyroclastic flows] and 1707 [explosive; mud flows]. In this last, whilst Cardinal Francis Pignatelli, with the clergy and people, devoutly followed the shrine of St. Januarius in procession to a chapel at the foot of Mount Vesuvius, the fiery eruption ceased, the mist, which before was so thick that no one could see another at the distance of three yards, was scattered, and at night the stars appeared in the sky. (Butler, "Lives of the Fathers")

Sir William Hamilton, in his 1767 monograph, *Campi Phlegraei*, described not only the eruption but the subsequent display of St. Januarius's head to avert the disaster:

> The noise and smell of sulphur increasing, we removed from our villa to Naples and I thought it proper, as I passed by Portici, to inform the court of what I had seen and humbly offered my opinion that His Sicilian Majesty (Ferdinand IV) should leave the neighborhood of the threatening mountain. However the court did not leave Portici till about twelve of the clock, when the lava was very near. I observed in my way to Naples, which was in less than two hours after I had left the mountain, that the lava had actually covered three miles of the very road through which we had retreated [the 1767 eruption consisted of two lava flows moving towards San Salvatore and Torre Annunziata; there were also vigorous explosions at the end of the eruption]. . . .
>
> The confusion at Naples this night cannot be described; His Sicilian Majesty's hasty retreat from Portici added to the alarm; all the churches were opened and filled; the streets were thronged with processions of saints. I shall avoid entering upon a description of the various ceremonies that were performed in this capital to quell the fury of the turbulent mountain. . . .
>
> Thursday, about ten of the clock in the morning the same thundering noise began again, but with more violence than in the preceding days. The oldest men declared that they had never heard the like, and indeed it was very alarming; we were in expectation every moment of some dire calamity. The ashes, or rather small cinders showered down so fast that the people in the streets were obliged to use umbrellas or flap their hats, these ashes being very offensive to the eyes. The tops of the houses and the balconies were covered above an inch thick with these cinders. Ships at sea twenty leagues [36 kilometers] from Naples were also covered with them, to the great astonishment of the sailors. In the midst of these horrors the mob, growing

tumultuous and impatient, obliged the Cardinal to bring out the head of St. Januarius and go with it in procession towards Vesuvius; and it is well attested here that the eruption ceased the moment the Saint came in sight of the mountain. (Hamilton, *Campi Phlegraei*)

Farther south in Sicily is Mount Etna, Europe's largest active volcano, with an elevation of 3,326 meters. Sicily is also the home of St. Agatha, well known for her ability to deal with volcanic eruptions. She was martyred in Catania, at the base of Mount Etna, where she was mutilated in prison. About A.D. 250, St. Peter cured her of the mutilation while she was in prison, but shortly thereafter she died. According to the legends, her veil has protected Catania from eruptions of Mount Etna. Von Waltershausen described one of these events: "During the eruption of A.D. 252, lava approached Catania and its inhabitants rushed to the tomb of St. Agatha, who had been martyred the year before, and carried her veil to the flow front. It was claimed at the time that the lava was immediately halted" (Von Waltershausen, *Der Ätna*).

Since that time, the veil has been tested on several occasions. In 1669 it was suggested that the veil prevented the complete destruction of Catania; the veil was brought out after part of the city had been destroyed, whereupon the lava changed course, flowed into the sea, and formed a huge breakwater. The veil was still in use in 1886, when Nicolisi was threatened—the local townspeople had used their patron saints to no avail, but when the Bishop of Catania arrived with the garment, the flow stopped. More towns have been destroyed than saved, however, despite the use of such intercessions.

In his study of Mount Etna, David Chester notes that even today, appealing to a deity for help remains a common response. The intensity of these appeals has been softened somewhat, for there is public relief—such as temporary housing and insurance.

The Making of a Myth

The youthful volcanoes of New Zealand's North Island have been fertile grounds for myths. The Tongariro Volcanoes are sacred to the Maori People and have their own legend: Taranaki (formerly Mount Egmont, 125 kilometers to the west of Ruapehu) once stood in the space between Tongariro and Ruapehu (two of the Tongariro Volcanoes). One day Tongariro seduced Ruapehu while Taranaki was absent. Upon his return Taranaki was so enraged that he stormed off, carving the Whanganui River valley. Stopping as the sun went down, he looked back at his wife, Ruapehu, and does so to this day. The fire of their eruptions is credited to Ngatoro-i-rangi of the Awara people.

The short but violent eruption of Tarawera Volcano in 1886 was well covered by the world press, and a detailed history has been written by Ron Keam, a professor of physics at Auckland University. The eruption has been a passion

Fig. 10-8. Artist's depiction of the Phantom Canoe, a myth associated with the 1886 eruption of Tarawera Volcano, North Island, New Zealand. (Source: postcard, artist unknown)

with Professor Keam, who has incorporated every historical, mythological, and geological observation about it into this amazing tome. One of the chapters is devoted to careful detective work on the legend of the Phantom Canoe, which was not an ancient story passed down from generation to generation by storytellers, but rather a modern tale promoted in the newspapers.

The story of the Phantom Canoe began in the Maori community of Ohinemeta, a center of activity on Lake Rotorua for generations. The people had chosen to live there because of the lake's obvious natural advantages, hot water for cooking, washing, bathing, and recreation. However, there were also dangers, including fatal scalding of the unwary. Carved posts in the shallow water offshore testified to the natural subsidence of the lake, which happened generations before the eruption.

On May 30, 1886, some tourists, who were being led by the guide Sophia, saw a large war canoe manned by thirteen paddlers. The tour's Maori rowers whistled and called to the strangers, but there was no reply. It was a clear day, and even the flash of paddles was visible (fig. 10-8). Eventually, the tourist boat turned in to the shore and lost sight of the canoe. The same day, the lake waters began to surge and creeks dried up—probable hydrologic precursors to the eruption.

Upon hearing of the Phantom Canoe, the New Zealand press picked up the story and enhanced it as a supernatural warning of Tarawera's eruption. Sophia lived and worked for another twenty-five years afterward, telling the story perhaps several hundred times. Gradually, the legend gained more supernatural elements, perhaps as a result of failing memory, tourists' expectations, or the background of a more traditional culture. And so the story became the myth of the Phantom Canoe. There are many possible explanations for it. The village was in an unsettled mood, possibly because of deaths caused by disease or even by an increase in alcoholism, and the eruption's precursor geophysical phenomena, such as steam blasts, probably added to the unease. It was also the day of a village chief's funeral, and important people in this culture were buried secretly in caves on the slopes of Tarawera. The canoe could have been a burial party going to the volcano.

Because of its close association with the eruption, the Phantom Canoe caught the public's imagination, and some people believe that it portended the eruption. Many Maori people, however, saw no reason to give a supernatural explanation of the sighting as a premonition of the eruption. But newspapers and tour guides viewed it as an opportunity for a great story—a myth in the making.

The Emergence of Volcanology as a Science

In the late sixteenth century, *volcan*, without the Italian ending, was the preferred word in the English-speaking world to refer to volcanoes, but by 1613 it had been accepted in English as *volcano*, along with the incorrect descriptive definition of "burning mountain." The words *fire* and *burn*, which refer to destruction by combustion, do not apply to volcanoes—they produce hot lava that can cause fire, but volcanic explosions and lava are not the products of combustion.

The adjective *volcanic* entered English in 1774, thereby firmly implanting its companion, *volcano*, in the language. The word *volcanology* to describe the science of volcanoes came into being in 1886, and *volcanologist* appeared in 1890, although volcanology as a distinct organized and financed science did not appear until after 1902. It is young as a geological science but is old as man's mythology.

Volcanology began to disengage itself from mythological ideas during the nineteenth century and become a scientific study with the eruption of Krakatau in 1883, which can be regarded as the preface to modern volcanology. Twenty years later, the 1902–1903 eruption of Mount Pelée underscored the need for more knowledge about volcanoes so firmly that it can be regarded as the establishment of volcanology as a distinct science.

The eruptions of Krakatau and Mount Pelée alone were responsible for the deaths of nearly 66,000 people, which shocked the world and accelerated funding for the study of eruptions. It is a sad commentary, but the amount of research on volcanoes at the turn of the century depended in large part on the size of the disaster in terms of total deaths. Modern volcano research is also heavily weighted toward certain areas, for example, those of greatest funding or national interest, but within the last decade, much of it has been aimed at understanding the processes by which eruptions occur. Such pure research has had great practical value in mitigating volcanic hazards. The quality of the research and therefore its value as an instrument of learning depends upon the scientists who do it and the state of available technology at the time of the study. For example, an estimated 92,000 people died in the 1815 eruption of Tambora (about 80,000 of them by starvation), but science, technology, transportation, and communications had not reached the point where help or meaningful research could occur.

By the time of the Krakatauan and Peléean eruptions, technology and scientific approaches had reached a stage of development that allowed more in-depth research into the causes and results of volcanism. Following the eruption of Krakatau, for example, the Dutch and British governments funded expeditions to the island and appointed scientists to study the eruption to assess damage in surrounding areas, and to examine global effects upon the environment, particularly upon climate.

Science and Modern Ritual

We have now reached the point in volcanology whereby the events of past eruptions can be explained in greater detail than was provided by the people who witnessed them. This includes the eruption of Mount Vesuvius in A.D. 79 that destroyed Pompeii, Herculaneum, other villages, and many villas; the eruption of Tambora in 1815; and the eruption of Krakatau in 1883. But although we can explain some of the natural phenomena of volcanic eruptions, and therefore help save lives by using this knowledge, many people of the modern world believe that prayer, divine intervention, and even sacrifice, can turn back lava flows and prevent destructive eruptions from occurring.

The confluence of the modern world and elemental reactions to a volcanic eruption is illustrated by the eruption of Agung in Bali (see chap. 9). The eruption began while worshippers were praying on the mountain at the great Temple of Besakih. The Bali Department of Foreign Affairs officially explained that the eruption was a sign from the mountain gods. Because pamphlets from the tourist department had already been printed, the governor ordered the religious ceremony known as the Great Rite to proceed as planned, despite the risk of affronting the mountain deities.

During the rituals of Agung, there were many offerings, such as cattle, pigs, and poultry, that were dropped into a nearby lake to please the water spirits and offerings that were made on the slopes of the volcano. Despite the offerings, Agung erupted; 1,550 people lost their lives, 85,000 were left homeless, and a third of Bali was covered by ash. Because the religious rites were not stopped when the volcano began erupting, many people refused to accept relief supplies sent to Bali from Canada, the United States, the Philippines, Thailand, and France. In this day of scientific knowledge about the earth, the people believed that lives were lost due to the displeasure of the gods, and therefore they chose to suffer unnecessarily so as to exorcise their guilt.

SUPERSTITIONS do not die, they are transmitted from generation to generation, and if not transmitted, they are born anew in a terrified human being who does not have the knowledge to understand nature's terror. But there are compensations. We start our next chapter with ski trips in New Zealand, and then turn to the uses of volcanic rocks, some right in your own household.

References

Boorstin, D. *The Creators: A History of Heroes of the Imagination*. New York: Random House, 1992.

Butler, Alban. "The Lives of the Fathers, Martyrs, and other Principal Saints," 1771–1773. Quoted in D. Seward, *Naples: A Traveler's Companion*. New York: Atheneum, 1986, p. 125. Annotations in brackets are the authors'.

Chester, D. K., A. M. Duncan, J. E. Guest, and C.R.J. Kilburn. *Mount Etna: The Anatomy of a Volcano*. Stanford: Stanford University Press, 1985.

Doumas, C. G. *Thera: Pompeii of the Ancient Aegean*. New York: Thames and Hudson, 1983.

Hamilton, William. *Campi Phlegraei*, 1767. Quoted in D. Seward, *Naples: A Traveler's Companion*. New York: Atheneum, 1986, pp. 131–32. Interpolations in brackets are the authors'.

Keam, R. F. *Tarawera: The Volcanic Eruption of 10 June 1886*. Auckland, New Zealand: published by the author, 1988.

Luce, J. V. *The End of Atlantis*. Paladin, Frogmore, St. Albans, 1974.

Mavor, James. *Voyage to Atlantis*. New York: Putnam, 1969.

Thorarinsson, S. *The Eruption of Hekla—1947–1948. I. The Eruptions of Hekla in Historical Times: A Tephrochronological Study*. Reykjavik: Societa Scientarium Islandica, 1967.

Vitaliano, D. B. *Legends of the Earth: Their Geologic Origin*. Bloomington, Ind.: University of Indiana Press, 1973.

Von Waltershausen, S. W. *Der Ätna*. 2 vols. Leipzig: Engelman, 1880.

Volcanoes for Consumers

Frontispiece. Cinder cones, the most common type of volcano in the world, are quarried for construction materials and therefore are an endangered volcano species. The cinders, usually red oxidized basalt, are used for ornamental stones and pathways, and often are locally used for road construction. This photograph shows the simplicity of a typical quarrying operation at a cinder cone near Little Lake, California. All that is needed is a bulldozer (right) to load the cinders into trucks. Steps have been taken by environmental groups to prevent the rapid destruction of cinder cones. (Photo: Richard V. Fisher)

Fig. 11-1. Heat that leaks from cooling magma chambers encounters groundwater to make steaming pools and mudpots. Here an exuberantly steaming crevasse near mudpools at Rotorua, New Zealand, shows an eerie, primordial scene. (Photo: Richard V. Fisher)

Skiing and Hot Baths

Not so long ago geologically, numerous volcanic eruptions created the landscape on which Auckland, New Zealand, is situated. Volcanoes kept changing the landscape over a period of sixty thousand years, the last eruption witnessed by Maoris only six hundred years ago. The city of Auckland is built on and around forty-eight cinder cones, on lava flows, and, in former marshy areas, on tuff rings. Many of the cinder cones and tuff rings have been destroyed—quarried for their valuable cinders, which are used for roads, walls, and buildings—but some have been preserved as parks. In ancient times, the high ground of Auckland's cinder cones provided protection to the native population—the flat terraces on which fortifications were built still exist today on several of the cones. The ancient Maoris also took advantage, as we do today, of Auckland's harbors, which were formed and protected by coalesced volcanic slopes. Some of the smaller yacht moorings are actually within the craters of breached tuff rings.

But many people regard Auckland as the gateway to a volcano winter wonderland. It rarely snows in Auckland and none of its volcanoes are over 260 meters high, so during winter, skiers travel southward to Tongariro National Park with its many high-standing, snow-covered, active volcanoes. The highest volcano in the park, Mount Ruapehu, has one of the finest ski fields, but the volcano occasionally erupts through its crater lake, producing lahars that sometimes damage the ski area. This happened in September 1995, when Ruapehu reawakened and one hundred thousand people within a 100-kilometer radius went on alert. On September 23, Mount Ruapehu erupted ash, rocks, and steam into the air and sent three lahars pouring down the sides of the volcano. There is always the possibility that a major eruption can occur at Ruapehu, but skiers and other tourists still come by the droves.

Ruapehu lies within the Taupo Volcanic Zone, a volcanic field where there have been numerous eruptions over the last twenty thousand years. Such volcanic fields are constructed on crustal rocks overlying hot magma bodies that feed the volcanoes. And though eruptions may eventually stop, tens to hundreds of thousands of years are needed for the magma bodies to completely cool. Over the years, water from snow and rain percolates downward through cracks and eventually encounters large areas of magma-heated rock that is hot enough to make steam, which then rises to the surface to make geysers, boiling mud pots, and steaming crevasses (fig 11-1).

Some people claim that a hot mud bath is far more beneficial to the mind and spirit than the generation of heat or electricity. Should a skier need soothing after the rigors of the slopes, the road north from Ruapehu, especially as it passes through the towns of Taupo and Rotorua, is lined with numerous spas where relaxation is the reward. More than eight hundred wells have been drilled in Rotorua and though many are quite smelly, their lure is hot water—elegant, traditional hotels with baths, large community swimming pools, and rows of motels, each with natural hot-water tubs and hot-water-heated rooms. These accommodations and the region's mud pots and steaming hot springs make Rotorua a holiday destination. The city offers a unique golf course with traps of boiling water as well as sand traps—the devil's own golf course. Neighboring Taupo also has a selection of pools to choose from, including one that is divided into sections, with one side kept at 36°C and the other at 40°C.

Although many people shy away from residing in a live volcanic geothermal region, there are benefits. In the towns, villages, and farms of the region, some people have their own geothermal well and consequently no heating bills. One of the main drawbacks is that some geothermal springs exude the stench of various sulfur compounds. Another drawback is an occasional explosion. Some bathhouses have been destroyed by steam explosions, and in one case, steam explosions blasted craters that swallowed a trailer and part of a house.

Construction Materials and Kitty Litter

A heavy rain of volcanic ash can make it very difficult to breathe and may cause roofs to collapse, but long after the bad effects pass, volcanic ash becomes a natural resource for construction and for industry. The same applies to many lavas and tuffs that can be cut into blocks and used as building stone. Very fine-grained volcanic ash, usually hundreds of kilometers downwind from its source, is widely used as polishing compounds—including in toothpaste and in household scrubbing compounds.

One of the least appreciated volcanic building materials—concrete—profoundly influenced the course of western civilization and is the cornerstone of modern society. Ancient Egyptians used a cement mortar made with gypsum to help construct the great pyramids at Giza, but gypsum cement eventually dissolves. A revolution in concrete-making by the early Romans helped indirectly ensure that Rome would become the dominant Mediterranean and European power. About 150 B.C., the Romans discovered that granular volcanic ash mixed with lime cement can set up under water and create a water-resistant concrete with superior strength. They called the volcanic ash *pozzuolana*, because it was quarried near Pozzuoli, Italy. Pozzuolana was also mined in other volcanic centers in the Aegean world, most notably at Thera (Santorini) (see chap. 10), where it was called Santorini earth. This "hydraulic

Fig. 11-2. A monumental building in any century, the Pantheon was constructed by the Romans two thousand years ago and has been in use for almost as long. At the time it was a spectacular architectural achievement made possible by concrete developed by the Romans who dared to make a building with a circular unsupported dome 44.3 meters in diameter. (Photo: Grant Heiken)

cement" allowed the Romans to build monumental buildings, extensive water works, viaducts, harbors, underwater pilings, and lighthouses. Pozzuolana is used in concrete structures to this day and can be emplaced under water through pressurized hoses.

Two thousand years later, many examples of Roman concrete construction still exist—the Pantheon, the Colosseum, Hadrian's mausoleum, and various Roman baths. The Pantheon in Rome (fig. 11-2), completed in A.D. 124 with a circular dome 44.3 meters in diameter, has survived nearly intact and is still in use. Pozzuolana cement allowed construction of the still-standing Pont du Gard, an aqueduct across the Gard River, near Nimes, France. Part of a 39-kilometer-long aqueduct completed in 18 B.C., the Pont du Gard stands 48 meters high and has a top level that supports a mortared water conduit.

Pozzuolana cement, which was used mainly where volcanic ash was easily available, also ensured that the Roman roads would be long lasting. The road

Fig. 11-3. A house in east central China made of welded ignimbrite sawed into slabs. Pillars, cornices, and tongue-in-groove walls and porch are all cut from the ignimbrite. Note fiamme (see chap. 5) in lintel. (Photo: Richard V. Fisher)

beds, commonly 1.25 meters thick, were layered with sand and gravel at the base, pozzuolana in the middle, and stone blocks at the surface. Main roads through the empire totaled nearly 87,000 kilometers with another 322,000 kilometers of connecting roads. Though bumpy, Roman roads lasted thirty to forty years without repair, greatly exceeding the lifetime of our modern highways. The routes of some of them still exist after two thousand years, and the original construction materials can still be found in some places along their way.

Tuff, with its particles bonded by natural welding or with cement, makes excellent building blocks. The most common volcanic building stone is ignimbrite, which is usually sawn from a quarry face and then cut into blocks by hand or by power saws. These blocks have enough strength for moderately high buildings, stone walls, and other structures, and they are lightweight, are resistant to weathering, and are good insulators—better than all other natural building stones.

The partly welded and cemented Campanian Ignimbrite extends over thousands of square kilometers in the region around Naples (see chap. 5). The pyroclastic flow that formed the Campanian Ignimbrite left the land barren 35,000 years ago, but the deposit is a historic architectural blessing, along with the less voluminous Neapolitan Yellow Tuff, which followed later. Ancient

Fig. 11-4. From prehistoric to modern times, homes have been carved into readily excavated tuff, which is an excellent insulator, holding heat in cold winters and remaining cool
in hot weather. In modern Italy, tourist shops at San Angelo, on the island of Ischia, are
carved into tuff layers. (Photo: Richard V. Fisher)

Etruscans, Greeks, and Romans and modern Italians have easily quarried and
cut the abundant ignimbrite into blocks and bricks to construct buildings, shelters, roads, walls, aqueducts, and fences for more than two thousand years.
Many of the buildings constructed of tuff in central Naples during medieval
times are still in use today. The durability of ignimbrite is remarkable, as
shown by Armenian architectural monuments, which are still in good condition one thousand to fifteen hundred years after their construction. Vast tracts
of land in Armenia, Georgia (formerly part of the USSR), Turkey, eastern
China, and much of Latin America are underlaid by ignimbrite that could be
used as building materials for houses, monuments, and roads (fig. 11-3).

Slightly welded tuffs, which are easy to dig holes and caverns into, have
been used as cave dwellings for centuries. In many places, these tuff caves are
still used for storage and animal shelters. Picturesque examples of this construction type are located in Turkey, where dwellings were carved into cliff
faces because the ignimbrite was easy to excavate, structurally sound, an excellent insulator, and fireproof, and the caves were easily defended. On the
island of Ischia, Italy, modern tourist shops and homes occupy rooms in excavated tuff (fig. 11-4).

Fig. 11-5. Sophisticated carvings have been cut into spires of tuff that house shrines and living quarters. They have been in use for centuries on the central Anatolia plateau of Turkey. (Photo: Courtesy of Turkish Tourism Bureau)

The central Anatolian Plateau of Turkey is partly covered by poorly welded ignimbrites erupted sporadically throughout the last five million years. These deposits exhibit typical "tent rock"—closely-spaced cone- or tent-shaped land forms (fig. 11-5). These tuffs are good insulators against the bitter winter cold of the plateau, and caves have been carved into the tuffs continuously since Roman times. Byzantine churches, monasteries, chapels, and hermitages remain today and are now the focus of an active tourist trade. During the Arab invasions of Byzantium, these clerical communities were expanded into fortified underground cities, some consisting of as many as ten levels connected by tunnels. The city remnants are now occupied by farmers who have built stone houses across the cave entrances—with stone blocks carved from the same tuff deposits. With a little imagination, it might be assumed that the spires and mosques of the Ottoman empire were inspired by the distinctive, naturally eroded shapes of the ignimbrite. In Goreme, homes carved into the rock as long ago as 4,000 B.C. are well preserved. And during the eighth and ninth centuries, more Byzantine churches, monasteries, and hermitages were excavated. In nearby Georgia, the twelfth-century city of Vardzia, also quarried in tuff, has five hundred rooms and apartments, a chapel, banqueting halls, and stables. This type of construction did not withstand later Arab invasions but has resisted earthquakes and storms through the centuries.

Native Americans in the volcanic lands from northern Mexico to the southern tip of South America had few cave dwellings carved in tuff, but the use of shaped blocks of tuff was common in some areas. The White City of Arequipa, southern Peru has many elegant colonial buildings built of white, easily cut, lightweight, nonwelded ignimbrite blocks. Modern downtown Guadalajara, Mexico, also has many buildings, including a magnificent cathedral, built of blocks from the ignimbrite that underlies the city (fig. 11-6). Recent reconstruction of the downtown area required that traffic-free streets and the modern buildings be faced with slabs of the same ignimbrite, creating one of the most beautiful large cities of Mexico.

People in the Naples region of southern Italy have utilized volcanic rock for centuries. The Greeks carved a tunnel during the sixth century B.C. within twelve thousand-year-old yellow tuff at Cuma for religious purposes associated with the Sibyl, a prophetess. The passage, still in excellent condition after 2,600 years, is 131 meters long, 2.5 meters wide, and 5 meters high and still carries a mystical charm. The end of the tunnel has a slightly depressed floor and three small niches that enlarge the end of the tunnel where the Sibyl presented her oracles (fig. 11-7).

Many of the palaces, cathedrals, large commercial buildings, and apartment houses of Naples are built with blocks of yellow tuff—a fine-grained volcanic ash deposit consisting of glass, crystals, and clay and erupted from many different hydrovolcanic eruptions twelve thousand years ago (fig. 11-8). The yellow tuff is not the most beautiful of stones and is usually covered over with plaster. Hard, welded tuff—*piperno*—is used for uncovered columns and building trim, with its probable source being the older Campanian Ignimbrite (see chap. 5). Carved blocks of basaltic lava from Vesuvius have been used to pave some cobblestone streets in Naples. Such stones are exceedingly resistant to deterioration, even with the heaviest truck traffic.

Rock quarries were open-pit mines in the early years of Naples, but because the quarries destroyed too much fertile land, quarrying was shifted underground, creating vast caves and catacombs. In modern times, the use of giant earth-moving equipment has allowed open-pit quarrying to resume. Beyond

Fig. 11-6. Ignimbrite has the properties of being lightweight, with relative high strength to resist crushing under heavy loads, and is easy to cut and shape. This column is fashioned from ignimbrite in modern-day Guadalajara, Mexico, which has many magnificent buildings of the stone. (Photo: Grant Heiken)

the hills north of Naples on the Volturno Plain there are gigantic open-pit quarries that have been recently enlisted to solve the problem of garbage disposal—many of them are being filled with millions of tons of refuse and then topped with soil to reclaim valuable farmland.

For many years, cinder cones (see chaps. 3 and 4) around the world have been mined as a source of important construction materials. The mounds of easily dug, loose cinders can be loaded onto trucks with only minor preparation, which consists of sizing the cinders through sieves. Cinders are lightweight, yet strong, having rough surfaces that do not easily slide past one another like many crushed rocks used as road metal. The finer-grained cinders are used for sanding icy roads. Many cinder cones are composed of red cinders

Fig. 11-7. This triangle-shaped tunnel with skylights, 131 meters long, 2.5 meters wide, and 5 meters high, was excavated by Greeks during the sixth century B.C. for religious purposes. It is in excellent condition after 2,600 years. At the end of the tunnel is a slightly depressed floor and three small niches that enlarge the end of the tunnel where the Sibyl presented her oracles. (Photo: Richard V. Fisher)

that impart a distinctive color to roads, railroad beds, and athletic tracks. Because they are so cheap to mine, cinder cones are disappearing rapidly, but counterefforts to save them are being established.

Pumice is also commercially important. Pumice lumps—sponge-like, very lightweight chunks of volcanic glass—delight adults and children alike as "rock that can float"; hundreds of pictures have been taken of people holding huge boulders of pumice in their arms without effort. Pumice was used by

Fig. 11-8. *Top*: Workers load a truck with ignimbrite blocks from a quarry near Naples. The cut bricks are used to construct buildings and walls. *Bottom*: Santa Chiara Church in central Naples was constructed at the end of the thirteenth century out of cut blocks of Campanian Ignimbrite and yellow tuff. (Photos: Richard V. Fisher)

Mediterranean peoples in the past as an exfoliant scrub to remove dead skin or calluses and also as an abrasive cleaner, and it is still used as such today. The discovery of pumice in some archaeological excavations led to the conclusion that the ancient inhabitants of those places had left their land and villages because of fallout from a volcanic eruption, until it was discovered that the pumice was used for cosmetic purposes.

Today, there are industries in which pumice is mined and crushed or sieved to pea-size particles and then pressed and cemented in molds to form "cinder" blocks. The blocks are strong, lightweight, and of outstanding insulation value and therefore excellent for house construction. One such industry is at Laacher See, Germany, near Mainz, and the blocks are used in many places in the country where wood is at a premium. Modern pumice mines are controversial because the ground is stripped by huge earth movers that excavate large open-pit quarries. In Germany, the companies must refill the quarries in most instances and make them suitable for growing local crops. In the United States there are constant problems with quarry owners who are reluctant to restore the land. Although the production of building blocks is an active business, the most common commercial use for large quantities of pumice in the United States is to create "stone-washed" jeans.

Of historical significance is obsidian, the shiny black volcanic glass formed in silica-rich lava flows. Obsidian can be the same composition as pumice, but it is much denser because it lacks the millions of gas bubbles contained in pumice. Early in history, mankind discovered that the edges of obsidian are exceedingly sharp. It is easily flaked and so was fashioned into arrowheads and knives used in hunting and in war. Because of its utility, obsidian has been quarried and traded on all continents. Today, even in advanced cultures, obsidian is still fashioned into cutting tools; delicate flakes of obsidian are sometimes used in eye surgery.

Volcanic ash has important uses even after it has been chemically altered by reacting with water over time. The drilling industry (oil, gas, water, mining) uses tons of bentonite, a clay produced from volcanic ash, which is pumped into the ground as a slurry that performs a variety of functions, such as easing transport of ground-up rock to the surface. Bentonite is also used as a filler in adhesives, ceramics, bread, and cheap ice-milk products. There's even something for cat lovers—much of the cat litter sold commercially is bentonite, which easily absorbs moisture. Entrepreneurs take note: there is a very high profit margin on bentonite sold for kitty litter!

The Grand Tour

Mount Vesuvius is to the Naples region as Mount Fuji is to Japan—both are exciting and dramatic symbols for a nation. One difference is that Vesuvius has been intermittently active through two thousand years of continuous his-

tory, and with the legacy of Pompeii and Herculaneum, serves as a grim re-
minder of volcanic power. However, "[l]et's take a positive view. The mountain is an emblem of all the forms of wholesale death: the deluge, the great
conflagration (sterminator Vesevo, as the great poet was to say), but also of
survival, of human persistence. In this instance, nature run amok also makes
culture, makes artifacts, by murdering, petrifying history. In such disasters
there is much to appreciate" (the Cavaliere [Lord Hamilton]; Sontag, *The Volcano Lover*, p. 112).

Vesuvius was an important destination for travelers on the Grand Tour of
the eighteenth and nineteenth centuries. From 1734 to 1860, sunny Naples,
under the rule of the Neapolitan Bourbons, was a center of art and music and
of excursions to Vesuvius. And it was in those years that the buried ruins of
Pompeii were discovered. For thirty-six years during this period, Sir William
Hamilton was Britain's ambassador to Naples. He was a great amateur naturalist and received well-deserved praise for his publications on Vesuvius's eruptions between 1767 and 1794.

Writers and wealthy visitors promoted Naples and its volcano as one of the
most romantic stops on the Grand Tour. One enchanted tourist was Johan
Wolfgang Goethe, the philosopher, writer, and amateur naturalist who wrote
Italian Journey. Goethe climbed Vesuvius several times and once nearly lost
his life—his curiosity led him to the rim of the active crater, where he nearly
fell. But this near-fatal escapade did not lessen his curiosity, and he made his
third visit on March 20, 1787:

> The news that another emission of lava had just occurred, invisible to Naples since
> it was flowing towards Ottaiano, tempted me to make a third visit to Vesuvius. On
> reaching the foot of the mountain, I had hardly jumped down from my two-wheeled,
> one-horse vehicle before the two guides who had accompanied us the last time ap-
> peared on the scene and I hired them both.
>
> When we reached the cone, the elder one stayed with our coats and provisions
> while the younger one followed me. We bravely made our way towards the enor-
> mous cloud of steam which was issuing from a point halfway below the mouth of the
> cone. Having reached it, we descended carefully along its edge. The sky was clear
> and at last, through the turbulent clouds of steam, we saw the lava stream.
>
> It was only about ten feet wide, but the manner in which it flowed down the very
> gentle slope was most surprising. The lava on both sides of the stream cools as it
> moves, forming a channel. The lava on its bottom also cools, so that this channel is
> constantly being raised. The stream keeps steadily throwing off to right and left the
> scoria flowing on its surface. Gradually, two levels of considerable height are
> formed, between which the fiery stream continues to flow quietly like a mill brook.
> We walked along the foot of this embankment while the scoria kept steadily rolling
> down its sides. Occasionally there were gaps through which we could see the glow-
> ing mass from below. Further down, we were also able to observe it from above.
>
> Because of the bright sunshine, the glow of the lava was dulled. Only a little

smoke rose into the pure air. I felt a great desire to get near the place where the lava was issuing from the mountain. My guide assured me that this was safe, because the moment it comes forth, a flow forms a vaulted roof of cooled lava over itself, which he had often stood on. To have this experience, we again climbed up the mountain in order to approach the spot from the rear. Luckily, a gust of wind had cleared the air, though not entirely, for all around us puffs of hot vapor were emerging from thousands of fissures. By now we were actually standing on the lava crust, which lay twisted in coils like a soft mush, but it projected so far out that we could not see the lava gushing forth.

We tried to go half a dozen steps further, but the ground under our feet became hotter and hotter and a whirl of dense fumes darkened the sun and almost suffocated us. The guide who was walking in front turned back, grabbed me, and we stole away from the hellish cauldron.

After refreshing our eyes with the view and our throats with wine, we wandered about observing other features of this peak of hell which towers up in the middle of paradise. I inspected some more volcanic flues and saw that they were lined up to the rim with pendent, tapering formations of some stalactitic matter. Thanks to the irregular shape of the flues, some of these deposits were in easy reach, and with the help of our sticks and some hooked appliances we managed to break off some pieces. At the lava dealer's I had already seen similar ones, listed as true lavas, so I felt happy at having made this discovery. They were a volcanic soot, precipitated from the hot vapors; the condensed minerals they contained were clearly visible.

A magnificent sunset and evening lent their delight to the return journey. However, I could feel how confusing such a tremendous contrast must be. The Terrible beside the Beautiful, the Beautiful beside the Terrible, cancel one another out and produce a feeling of indifference. The Neapolitan would certainly be a different creature if he did not feel himself wedged between God and the Devil. (Goethe, *Italian Journey*, pp. 214–15)

Goethe and other famous travelers such as Percy Bysshe Shelley were also overwhelmed by the remarkably well-preserved ruins of Pompeii. Another unlikely writer who was favorably impressed by the combination of Neapolitan life, the ruins of Pompeii, and the grim grandeur of Vesuvius was Mark Twain, who visited in 1867.

Today Mount Vesuvius is quiet, having erupted last in 1944, but it still lures tourists who encounter the reality of a Roman past—Pompeii, Herculaneum, and Vesuvius (fig. 11-9). Private cars, tour buses, and taxis pour into the parking lot at the trailhead to the summit of Vesuvius. The air is filled with the languages of the world as the usually under-dressed tourists, whipped by cool winds, pay a thousand lire here and a thousand lire there—tickets to walk the summit trail, rental of walking shoes, and a tourist bazaar at the summit, where postcards, videotapes, garish ashtrays, and soft drinks are sold. But Vesuvius is not always crowded with foreign tourists. It also charms local Neapolitans. On sunny holidays and weekends, crowds of local people arrive at the summit

Fig. 11-9. Vesuvius when it last erupted in 1944. Strombolian-style eruptions that construct cinder cones have replaced the Plinian-style eruptions that produced pyroclastic flows and surges that destroyed Pompeii and Herculaneum in A.D. 79. No one knows if or when Plinian activity will return. The city of Naples and the millions of people living around Vesuvius are at the mercy and whims of Vesuvius. (U.S. National Archives photo 80-G-54421)

dressed in their Sunday best, strolling up the rocky trail in patent-leather dress shoes or in high heels, in contrast with Goretex®-encased Americans and sun-reddened northern Europeans. In some instances, you will see a formal wedding entourage with a bride in white lace and a groom in formal black attire standing on the volcano's rim being photographed with the crater in the background. Neapolitans are proud of the volcanic grandeur that is theirs.

Modern Volcano Tourism

"Explore the primal terrain of the volcano goddess Pele, where dramatic lava flows meet the sea!" proclaims an excursion brochure for the island of Hawaii; only $49.00 by bus, and $124.00 by helicopter. Everyday, thousands of tourists sample the excitement of a volcanic eruption. Rental cars are ready seven days a week for the 210-kilometer round-trip visit from Hilo, which includes the distance to Hawaii National Park, the Chain of Craters Road to where lava is entering the sea, and a few short side trips. For a budget tourist, the car rental is about $40.00, the park entrance fee is $5.00, and then there is a $2.00 dona- tion to the visitor center or the museum at the observatory, $6.00 for a lunchtime snack at Volcano House, and at least $25.00 for a T-shirt and mis- cellaneous souvenirs—a minimum of $78.00 for nature's grand spectacle, unique for its adventure and accessibility. But bring $200. In Hawaii, volca- noes are a consumer item and an important source of income.

Volcano tourism in Hawaii today contrasts markedly with the leisurely vol- cano tourism of nineteenth-century Italy. Passage is fast, comfortable, and safe. Traveling to an exotic tropical island far out in the Pacific Ocean is highly romantic, but seeing an eruption and flowing lava is thrilling to the extreme. The power, the sensations, and the geology are overwhelming; all of one's senses are assaulted, and geology comes alive. But it is dangerous to the care- less—outpourings of lava from vents, lava flowing explosively into the sea, and steep-sided craters with crumbling sides can cause injury or death. Then there are the human susceptibilities to heat prostration, heart attacks, or ex- haustion. To the National Park Service, tourism means safety responsibility and accident liability. Safety is their watchword, and to their credit, out of 2.5 million visitors in 1993, only one tourist was killed. A superb safety record.

Mount St. Helens is also another tourist attraction and has gone through several phases as such. First it fumed, swelled, and erupted ash from its cone, and millions of onlookers inundated the area. The second phase was the deadly northward blast that devastated hundreds of acres of timberland (see chap. 1). This phase attracted hundreds of scientists who began dozens of scientific studies of unprecedented value to volcanic hazards mitigation. Some volcano- logical mysteries have been unraveled, and many new questions have been asked that remain to be answered. The third and current phase began with the projects to make the devastated area a national monument and to build public observatories, interpretive trails, and monuments that teach the public about volcanic eruptions. From Interstate 5 at Castle Rock, a new paved road takes the tourist directly into the heart of greatest devastation from the 1980 erup- tion. From Johnston Ridge, the crater and dome are visible through the broken rim of the volcano. The volcano can also be seen from the northeast and the south from other roads.

During the several years since the area was devastated by the avalanche and pyroclastic flows, it has become an important attraction in the Pacific Northwest. There were 3.4 million visitors in 1993, 17 percent of them from foreign countries. But income from visitors is difficult to assess because many are local. As an estimate, if an average of $100 per day was spent by each person for an average stay of two days, generated income was easily in the hundreds of millions of dollars. Most money is spent on lodging, meals, gasoline, and souvenirs, but there are some cottage industries dependent on Mount St. Helens' fame, for example, Christmas tree ornaments made by glassblowers using ash from the 1980 eruption.

Plans for economic benefits from a volcano do not always become a reality. Izalco Volcano, in western El Salvador, for example, was a new volcano in 1769 and grew to 1,900 meters over the last two hundred years. Frequent activity gave it its nickname, The Lighthouse of the Pacific. Izalco had become such a tourist attraction that the Salvadoran government planned to increase its access to tourists. They constructed a hotel and an observatory on the cone of nearby Cerro Verde, where eruptions could be watched in comfort. But suddenly in 1965, a few months before completion of the hotel, Izalco stopped erupting. The Hotel Montana has been empty for years, but there is now a restaurant where lunch is served overlooking an attractive though presently inactive volcano.

TOURS on volcanoes can be arranged for the economic benefits of the guides, or a business that is formed to transport tourists. But oil, diamonds and gold from volcanoes? There is a connection, as explained in the next chapter.

References

Duffield, W. A., J. H. Sass, and M. L. Sorey. Tapping the Earth's Natural Heat. U.S. Geological Survey Circular 1125. Washington, D.C. 1994.

Goethe, J. W. *Italian Journey [1786–1788]*. Translated by W. H. Auden and E. Mayer. London: Penguin Books, 1962.

Paliotti, Vittorio. *Vesuvius: A Fiery History*. Naples: Cura E. Turismo, 1981.

Sontag, Susan. *The Volcano Lover: A Romance*. New York: Farrar Straus & Giroux, 1992.

CHAPTER 12

Volcanic Treasures: Steam, Gold, and Diamonds

Frontispiece. Blue Lagoon, Iceland, captures the essence of geothermal phenomena. Steam chimneys in the background typify the generation of power. Steam can also be piped directly to private homes for heat. Bathers in the hot pool in the foreground illustrate the recreational aspects of geothermal resources. (Photo: Gudmundur E. Sigvaldason, Nordic Volcanological Institute, Reykjavik, Iceland)

Imagine a winter snow piling high outside the window, and you sitting in a steaming bath, reading, relaxing, and letting the cares of the world pass by, all for little or no cost because you live in an active volcanic area. Your book is illuminated thanks to geothermally generated electricity—no need for coal or petroleum. Your house is made of blocks of tuff quarried from a nearby volcano, perhaps the same which yielded the gold encircling your finger. In harmony with their natural grandeur, volcanoes can be rich repositories of building materials, metals, precious stones, and especially clean, economical, and renewable geothermal energy.

Steam

The shallow magmas that can trigger destructive volcanic eruptions are also the beneficial heat engines driving many of the world's geothermal systems. The hottest waters circulating in these systems are an environmentally benign source of electric power. Cooler, but still warm, waters are used for such varied purposes as growing fish, fruits, and vegetables; drying foodstuffs; and heating residential and commercial buildings. Among many additional applications for geothermal energy, certainly the most ancient is simple relaxation—escape from the stress of daily life—with a soothing soak in a hot-spring pool.

Every year, the hundreds of hot springs around Beppu City, which is in the active volcanic region of northwestern Kyushu Island, Japan, attract more than twelve million visitors. The pilgrims come to bathe or swim in hot pools, to be smeared with hot muds, or even to be partly buried in hot volcanic sands. One cannot leave Japan, and especially Beppu, without feeling clean, relaxed, and in the glow of good health.

Japan, part of an active volcanic arc (see chap. 2), boasts eight to ten thousand hot springs, most of which, like those of Beppu, are developed as thriving vacation spas. The electric-power potential of these thermal areas is clearly enormous, but tourism is seen as even more profitable, and the spa owners are reluctant to sell their valuable lands for power generation, which could disrupt the hot springs' natural flow regimes.

Italy is another country blessed with a wealth of geothermal features, most of them associated with young volcanic activity. The country has more than seventy *terme*, or thermal baths, which are visited by people from throughout

Fig. 12-1. Hippopotamus Pool, on the Italian island of Vulcano, attracts thousands of visitors annually to Vulcan's forge. Warmed by rising heat from a shallow body of magma, this feature is named for the muddy bathers who frequently crowd the wallow and its shores. (Photo: Grant Heiken)

Europe. Many Italian hot springs have been developed as clinical spas because of their reportedly curative powers, and they are as popular today as they were with the Romans two thousand years ago. It is ironic that many of the people who flock to these facilities regard trace amounts of chemicals in their drinking waters as pollution. Yet the hot springs commonly contain high levels of evil-smelling sulfur, arsenic, chlorine, and bromine compounds—the visitors not only bathe in these waters, they eagerly drink them!

A geothermal pool of slightly acidic mud near the seashore on the island of Vulcano (see chap. 10) is popular with tourists who wallow in it, then stand around its edge, allowing the mud on their bodies to dry. The unusual appearance of these encrusted visitors has earned this thermal feature the name Hippopotamus Pool (fig. 12-1).

The still cooling rhyolitic magma bodies beneath the geothermal fields of New Zealand (see chap. 11) have erupted 16,000 cubic kilometers of pyroclastic debris in the last two or three million years, yet many remain potent heat sources. The magmas themselves and their crystallized equivalents contain little water, but with initial temperatures of 650°–1300°C, they readily heat waters that have been stored or are circulating in the subsurface. The hottest of these waters, which are accessed by geothermal wells, can be "flashed" to steam to generate electricity. Their slightly cooler counterparts can be piped directly to heat homes and other buildings. New Zealand geothermal power stations can supply 7 percent of the country's electricity, but they are presently used mainly as back-up facilities; most of the country's electricity is furnished by hydroelectric power plants. However, when energy needs inevitably expand, it is estimated that the geothermal sources will supply four to five times the power currently generated.

During California's gold rush in the mid-1800s, travel-weary prospectors enjoyed the many hot springs of the volcanic region near the Napa Valley, north of San Francisco—a region later to become famous for its world-class wines. By the late 1870s, there was a bustling hot-spring resort at the northern Napa Valley town of Calistoga. An artificial geyser was created there in 1928 when a water well penetrated a shallow thermal aquifer (an especially permeable and fluid-rich geologic formation) that had a temperature of 110°C. Before the well could be cooled, it exploded, forming a small crater at the surface. In spite of this violent event, the geothermal system beneath Calistoga's hot springs is not only small, but also lukewarm when compared to the one at The Geysers geothermal field about 25 kilometers to the northwest.

Although geothermal electric power was first produced at Larderello, Italy, in 1904, The Geysers (actually a misnomer, since there are no natural geysers in the region), with an area of at least 150 square kilometers, is the world's largest productive geothermal system. Like the system at Larderello, The Geysers is somewhat unusual in being "vapor-dominated," meaning that the high-temperature, high-pressure steam can be extracted directly from the geothermal reservoir to spin the power-producing turbines. In the more common "liquid-dominated" systems, hot waters tapped by deep wells flash only partially to steam. That fraction is used to generate electricity; the residual hot waters are used for other purposes or are reinjected into the geothermal reservoir rocks.

The Geysers steam reservoir is situated in the very young Clear Lake volcanic field, the northernmost and youngest in a chain of such fields that extends to the southeast, beyond the city of San Francisco, for several hundred kilometers. The lavas and tuffs of the Clear Lake field were erupted as recently as ten thousand years ago, and many of the Clear Lake volcanoes are little affected by erosion. A huge, cooling *pluton*—the crystallized equivalent of a magma chamber—several kilometers beneath one of these volcanoes has been dated at 1–2.5 million years, still quite young in geologic terms. It is this pluton and even younger and probably still partially molten intrusives that supply the region's geothermal heat.

Exploitation of the geothermal energy at The Geysers began in 1924, when fractured sandstones at shallow depths were drilled to produce steam at a temperature of about 100°C. The steam was used to generate electricity for a nearby resort and outlying cottages. Three years later, the geologists

E. T. Allen and A. I. Day estimated that a zone about 40 kilometers long and 1 kilometer wide at The Geysers was underlain by hot magma and predicted that electricity could be generated here on a large scale, but twenty-six years passed before this concept was tested. In 1953, Magma Power Company (MPC) leased land at The Geysers, drilled for steam, and by 1960 was furnishing steam to Pacific Gas and Electric Company to generate 12 megawatts of electricity (at current per capita consumption in the United States, one megawatt can supply a community of one thousand people). After MPC merged with Union Oil Company of California and Thermal Power Company in 1967, the new geothermal alliance drilled hundreds of additional steam wells to depths as great as 4,000 meters. By 1987, the field was producing about 2,000 megawatts of electricity, more than sufficient for the needs of greater San Francisco. Although steam pressures at The Geysers began to decline shortly thereafter, the field has so far produced several billion dollars worth of power and will no doubt remain productive well into the next century.

A geothermal system requires four critical components for the large-scale generation of electrical power: (1) a magmatic heat source, commonly but not always associated with one or more volcanoes; (2) hot water or steam in the rocks above this heat source; (3) sufficient permeability in the rocks for these fluids to flow at high rates under production; and (4) a nearly impermeable "caprock," which maintains pressure in the system by inhibiting thermal-fluid escape (fig. 12-2). Not all of these requirements are satisfied in every volcanic area, for instance, the rocks may be very hot but too tightly sealed to permit the passage of hot water or steam. Conversely, water may abound, with the heat source too small or too tepid to support a vigorous geothermal system. Geothermal producers and researchers have made good progress in recent years, devising means to compensate for the lack of permeability or natural waters in many systems with good supplies of heat. In the first and still largely experimental case, permeability is enhanced by inducing new fractures from deep wells under extremely high pressures. In the second case, involving now routine procedures, fluids are replenished by injecting into selected wells, cool waters, which quickly reheat upon contact with the hot reservoir rocks.

Another common use for geothermal energy is the heating of homes and commercial buildings, not only offices and storage facilities, but also greenhouses, within which such delicate crops as tomatoes, strawberries, roses, and orchids can be grown readily in otherwise cold climates or seasons. Some of the best examples of this practice are in Iceland, where the plentiful geothermal waters are even piped into homes for bathing and food preparation, making conventional hot-water heaters unnecessary. Iceland's geothermal wealth arises from a happy geologic coincidence. Not only does the island lie astride the heat-rich, midoceanic ridge or extensional fracture zone separating

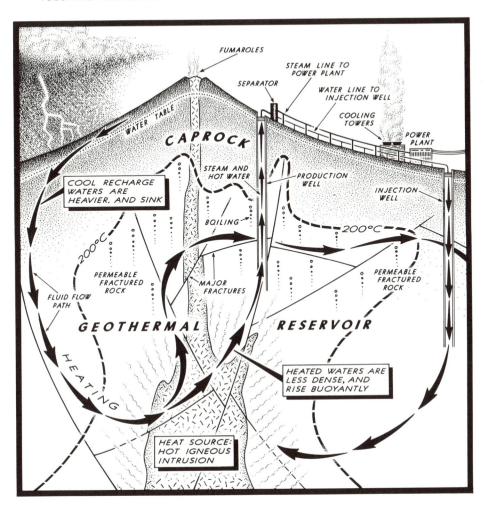

Fig. 12-2. Hot igneous intrusions beneath many volcanoes support large, high-temperature geothermal systems—clean, safe, and renewable sources of electrical power.

the American and Eurasian plates (see chap. 2), it also coincides with a huge, stationary hot spot in the earth, much like the one responsible for the Hawaiian Islands. As a result, a thick stack of mostly basaltic flows has built up several thousand meters from the seafloor. Associated, upwelling magmas have spawned numerous volcanoes while providing heat for literally hundreds of geothermal systems.

For both developed and developing countries, volcanoes can be an environmentally friendly source of geothermal power and warmth. El Salvador derives a third of its electrical power—and Costa Rica a tenth—from volcano-associated geothermal systems. Guatemala, Nicaragua, and Honduras are all starting geothermal programs. In the Philippines and Indonesia, large geothermal systems that can produce electricity are found around many of these countries' stratovolcanoes. As the earth's fossil and nuclear fuels are inevitably depleted, geothermal energy will become increasingly important as a clean, renewable source of electrical power far into the future.

Oil and Volcanoes Don't Mix . . . or Do They?

Most oil companies have traditionally viewed volcanoes and volcanic land as unfavorable targets for petroleum exploration. The rationale for this philosophy has been that the intense heat of volcanoes and their geothermal systems would be unfavorable for the generation and preservation of petroleum—that otherwise free-flowing oils would be useless tarry residues in such environments. While this is true in extreme cases, there is now evidence that under certain circumstances, volcanic terranes can be prime petroleum prospects and that geothermal systems can be instrumental in petroleum generation, migration, and entrapment.

The most dramatic clue to this phenomenon is afforded by the oil-bearing, newly deposited seafloor sediments associated with black smoker hydrothermal vents at midocean ridges (see chap. 4) in the northeast Pacific Ocean. Geologists from the USGS and Oregon State University have shown that the oils in these sediments were hydrothermally generated, nearly instantaneously, at temperatures of up to 315°C. This is twice the temperature considered the uppermost limit for oil generation under so-called "normal" conditions—that is, over long periods of time deep within a thick sedimentary rock sequence that increases in temperature about 30°C for every kilometer of depth. Geothermal oils have since been found in hot springs and hot-spring deposits at the Waiotapu geothermal system in New Zealand and at the edge of the huge caldera cluster of Yellowstone National Park (see chap. 5). In addition, several extinct high-temperature geothermal systems have been shown to contain abundant oils and tars directly associated with gold, silver, and mercury, suggesting that the organic compounds may have stimulated metal precipitation. The true economic value of geothermal oil on a worldwide basis remains to be established, but there is no doubt that high-temperature geothermal processes can play an important role in the genesis and evolution of petroleum.

Just as volcanic and associated geothermal processes have traditionally been viewed as hostile to the formation of oil reservoirs, so are volcanic rocks very uncommon as petroleum reservoir rocks. Imagine, then, the surprise of Shell

Oil Company geologists in 1954, when their deep oil-exploration well in eastern Nevada's Railroad Valley found abundant oil in ignimbrite (see chap. 5). Since then, ignimbrites in Nevada oil fields have produced 13 million barrels of oil worth about $156 million at today's prices. Needless to say, the region's formerly scorned ignimbrites are routinely targeted as excellent storage sites for rich pools of oil.

Volcanoes and Prospectors' Dreams

The high-energy geothermal systems generated by volcanoes are among the most efficient natural concentrators of valuable ore minerals. Some active, volcano-associated hot springs, for example at Waiotapu, precipitate oozes containing substantial amounts of gold (up to several ounces per ton). At Ngawha Hot Springs, also in New Zealand, mercury condenses from rising steam to drip from the eaves of bath houses. The geothermal systems beneath these metalliferous springs have the potential to deposit commercial concentrations of gold and other metals, including silver, copper, molybdenum, lead, and zinc. Although the active systems are almost always too hot to mine, their extinct, or "fossil" analogs in volcanic terranes account for a significant proportion of the world's mineral deposits. Instinctively realizing this, prospectors have long been attracted to regions of the earth where erosion has stripped extinct volcanoes to levels at which these deposits are conveniently exhumed.

The same factors critical for the creation and maintenance of a high-temperature geothermal system also favor formation of a mineral deposit: (1) a long-lived magmatic heat source; (2) high rock porosity and permeability; (3) a copious supply of water; and (4) a caprock to prevent ore-fluid escape. Hot waters heated by the magma circulate through and dissolve metals and other elements from the surrounding rocks. In some cases this process is abetted by superheated, metal-bearing waters expelled from the magma itself. The resulting ore fluids travel through the rocks until diminishing temperatures and pressures or dramatic changes in fluid or rock chemistry force them to drop their precious freight. In this way, and in repeated episodes, valuable metallic mineral deposits are formed.

The time required to build up these ores can be determined roughly either by dating associated veins of geothermal minerals or by estimating from present-day rates of metal deposition from hot volcanic waters or vapors. Using the first approach, geologists have calculated that development of a commercial precious-metal ore deposit takes at least several hundred thousand years. Results of the second approach, applied at Galeras Volcano in Colombia (see chap. 15), suggest that even quite large deposits may form in far less time. The volcanic plumes of Galeras contain trace amounts of gold, and in 1992 and 1993, the mountain erupted fragments of gold-bearing andesite. By combining

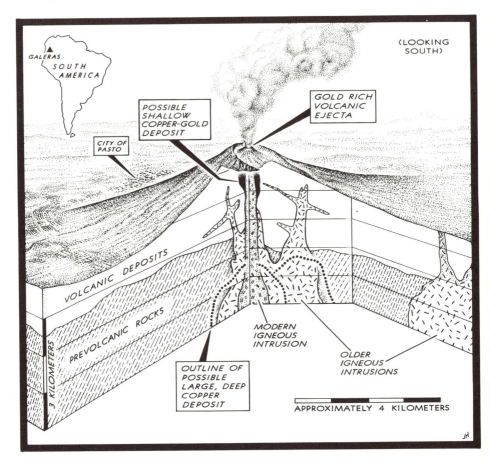

Fig. 12-3. Galeras Volcano, in addition to being a dangerous killer and one of the Decade Volcanoes, is also an efficient natural concentrator of gold, copper, and sulfur. Sulfurous, gold-bearing vapors emerge from Galeras, and volcanic rocks rich in the precious metal have been hurled from its depths during recent eruptions. Geologists speculate that these occurrences signal the shallow presence of rich copper and gold deposits beneath the mountain's crown. (Adapted from Goff et al. 1994)

the rate at which gold-bearing vapors now emerge from Galeras with other evidence, geologists at Los Alamos National Laboratory determined that a moderate-sized ore body containing 400,000 ounces of gold (currently worth $160 million) could form in only ten thousand years. They also speculated that deep within the volcano, rich gold and copper deposits may even now lie fully formed (fig. 12–3), perhaps one day to yield these metals for our descendants.

Volcanic Gemstones

In 1867, the course of history of southern Africa was dramatically altered when a small shiny rock was plucked from river gravels by a shepherd in the veld of west Griqualand. The "rock" was a large diamond, and the find lured a rush of prospectors, who eagerly sifted the gravels for more. By 1871, the gravels had been traced to the Kimberly area, where a nondescript hill called Colesburg Kopje was revealed to be the filled, eroded throat of a long-extinct, diamond-bearing volcano. The material filling this "pipe" was accordingly named *kimberlite*. For several years, a mine developed in the Kimberly pipe produced 90 percent of the world's diamonds. The rich supply was controlled by just one company—DeBeers Consolidated Mines, which is still the major company in the world's diamond trade. Within a few decades, what had been Colesburg Kopje was a deep hole extending more than 3 kilometers into the earth. Through the years, this mine and others have made what is now the Republic of South Africa the world's foremost diamond producer.

Volcanoes bring diamonds from deep within the earth, where unimaginable pressures squeeze carbon atoms into tightly bonded molecules to form the hardest of known natural substances. Gem-quality diamonds are not only lovely, with their incomparable internal brilliance, they are exceedingly rare; for example, whereas the grade of an average volcanic gold deposit is about 0.07 ounces per ton of mined rock (a little more than 2,000 parts per billion), that of a diamond deposit (excluding industrial-grade stones) may be as little as 7 parts per billion!

The richest diamond mines are in South Africa and Western Australia, but diamond-bearing kimberlite pipes and gravels have also been found in Botswana, Brazil, Canada, Siberia, and the United States. Diamonds were discovered in a kimberlite region that spans parts of Wyoming and Colorado in 1975, and this area, the United States' top producer, has since yielded stones weighing a total of 120,000 carats.

Kimberlite volcanoes are exclusively ancient features, having been active between about 0.1 and 3.2 billion years ago. All are now eroded down to their throats, and many are little more than weathered masses of bluish clay (for example, the famous "blue ground" of the South African deposits). Geologists believe that the kimberlite conduits once led upward to tuff rings, which were produced by what must have been spectacular hydrovolcanic eruptions (see chap. 4).

Studies of the chemical and physical properties of diamond-bearing kimberlite have revealed that the material was explosively emplaced by volatile-rich magmas propelled by a mixture of steam and carbon dioxide. These gaseous magmas, with their sparse cargo of diamonds, are believed to have originated in the earth's mantle at a depth of perhaps 200 kilometers and to have forced their way to the surface at velocities of 10 to 30 kilometers per hour. Once near the surface, violently escaping gases and volcanic ejecta (including diamonds) formed the signature tuff rings. It is clear that this phenomenon is exclusively an ancient one, induced by a formerly more energetic earth. No young kimberlite tuff rings have been discovered, nor are recent tuff rings of other compositions known to be diamond-bearing.

Diamonds have fostered many a fortune, mostly legitimate but sometimes the illicit yield of elaborate swindles. In 1872, five men bought 10 pounds of uncut diamonds, plus 50 pounds of garnets and other semiprecious gems from Arizona and planted them carefully in a sandstone outcrop in Wyoming (in the 1870s sandstone was believed to be a diamond host rock). The conspirators whispered of the findings to a few people in the mining industry, including the former Civil War general George B. McClellan. Intrigued by these reports, McClellan and his colleagues hired a prominent mining engineer to examine the outcrop. The engineer, who had never seen a diamond site, vouched for the value of the prospect. On this advice, the investors formed a mining company and bought the property for several hundred thousand dollars, the equivalent of millions today. Unfortunately the land belonged to the federal government, and by the time the new "owners" became aware of this fact, the villains had vanished.

Another extremely valuable volcanic gem is the red beryl, or "red emerald" of the Wah Wah Mountains of southern Utah in the United States. This claret-colored stone is even rarer than diamond, and a single, uncut crystal specimen recently sold for $200,000. The beryl occurs in a 22-million-year-old rhyolite flow, which also contains scattered topaz. Unlike diamond, the beryl formed near the surface at rather low pressures, although at temperatures probably in excess of 300°C. It is believed that the beautiful red crystals were deposited in fractures by beryllium-rich vapors degassing from the rhyolite flow shortly following its eruption.

THESE gems, precious metals, and other volcanic treasures have enriched peoples' lives for millennia. Even so, their value has been greatly surpassed by the agricultural bounty of rich volcanic soils, a topic we now explore.

References

"Diamonds: Wyoming's best friend." *Geotimes* 41, no. 2 (1996): 9–10.

Duffield, W. A., J. H. Sass, and M. L. Sorey. *Tapping the Earth's Natural Heat*. U.S. Geological Survey Circular 1125. Washington, D.C., 1994.

Gilbert, J. M., and C. F. Park, Jr. *The Geology of Ore Deposits*. New York: W. H. Freeman and Co., 1986.

Goff, F., et al. "Gold degassing and deposition at Galeras Volcano, Colombia." *GSA Today* 4 (1994): 244–47.

Rytuba, J. J., ed. *Active Geothermal Systems and Gold-Mercury Deposits in the So-*

noma-Clear Lake Volcanic Fields, California. Society of Economic Geologists Guidebook Series, Vol. 16. Ft. Collins, Colo., 1993.

Schalla, R. A., and E. H. Johnson, eds. *Oil Fields of the Great Basin.* Reno: Nevada Petroleum Society, 1994.

Sillitoe, R. H., and H. F. Bonham, Jr. "Volcanic landforms and ore deposits." *Economic Geology* 79 (1984): 1286–98.

Stone, C., ed. *Monograph on The Geysers Geothermal Field.* Geothermal Resources Council Special Report 17. Davis, Calif., 1992.

Wohletz, K., and G. Heiken. *Volcanology and Geothermal Energy.* Berkeley: University of California Press, 1992.

From Ashes Grow the Vines

Frontispiece. Vineyards have graced the slopes of Vesuvius since pre-Roman times when Greeks inhabited the shores of ancient Italy. The rich, well-drained volcanic soils on the flanks of the mountain supported not only grapevines but also citrus trees, vegetables, herbs, and flowers—agricultural bounty celebrated by Roman authors before and since the eruption of A.D. 79. And though much of this land is now occupied by cities, Vesuvius remains an attractive fertile garden.

On February 20, 1943, a fissure in a cornfield in Mexico opened up and an event occurred that is extremely rarely witnessed—the beginning of a volcano where none had existed before. The three closest witnesses to the birth of the volcano were the farm owner, Dionisio Pulido; his wife, Paula; and Demetrio Toral, a laborer from the village of Parícutin who worked for Pulido. Over the previous forty-five days, all of the local people, including Dionisio and Paula, had been experiencing earthquakes. Dionisio and Paula had been hearing thunderlike sounds from under the ground. At about 4 P.M. Paula was watching sheep about 100 meters from the initial wisps of smoke that began blowing out of the fissure. Dionisio was burning shrubbery about 175 meters away, and Toral was plowing the cornfield about 150 meters away. He had just completed a furrow and was turning around when the first fissure opened almost in the exact furrow he had just plowed. (This led some people to say later that Toral "plowed up the volcano.") Dionisio watched, stunned:

> [T]he ground swelled and raised itself 2 or 2½ m high, and a kind of smoke or fine dust—gray, like ashes—began to rise up in a portion of the crack. . . . Immediately more smoke began to rise, with a hiss or whistle, loud and continuous, and there was a smell of sulfur. I then became greatly frightened and tried to help unyoke one of the ox teams. I hardly knew what to do, so stunned was I before this, not knowing what to think or what to do and not able to find my wife or my son or my animals. Finally my wits returned and I recalled the sacred Señor de los Milagros, which was the church in San Juan Parangaricutiro and in a loud voice I cried, "Santo Señor de los Milagros, you brought me into this world—now save me from the dangers in which I am about to die"; I looked toward the fissure whence rose the smoke; and my fear for the first time disappeared. I ran to see if I could save my family and my companions and my oxen, but I did not see them and thought that they had taken the oxen to the spring for water. I saw that there was no longer any water in the spring, for it was near the fissure. . . . Then, very frightened, I mounted my mare and galloped to Parícutin, where I found my wife and son and friends awaiting, fearing that I might be dead and that they would never see me again. On the road to Parícutin I thought of my little animals, the yoke oxen, that were going to die in that flame and smoke, but upon arriving at my house I was happy to see that they were there. (quoted in Luhr and Simkin, *Parícutin*, p. 56)

The eruption started in a hole that Pulido had been trying to fill for years. The following day, February 21, a mound of cinders that erupted from the

Fig. 13-1. Unwanted fame came to Dionisio Pulido shown here with his nemesis, Parícutin Volcano in the background. Dionisio was a successful farmer until February 20, 1943, when an eruption began from a fissure in his cornfield in Mexico. Dionisio lost his farm to the ash, cinders, and lava flows that devastated the surrounding land. (Photo: Mary St. Albans, from the Smithsonian Institution)

fissure had built a volcano 10 meters high and rocks were being hurled from it into the sky (fig. 13-1). That was a mere beginning. During the next nine years, more than a billion cubic meters of volcanic ash and 700 million cubic meters of lava were erupted from Dionisio Pulido's cornfield, forever changing the lives of everyone in the five villages of Itzicuaro Valley (fig. 13-2). The cinder cone grew to a height of 424 meters before the volcanic activity stopped.

Parícutin Volcano became well known during the 1940s. It is now a large, inactive cinder cone that sits in Itzicuaro Valley, Michoacan, within the Trans-Mexican volcanic province. Thousands of acres of this highland valley were

Fig. 13-2. The boundaries of rich farmland fields of central Mexico appear by perspective to converge upon the eruption of Parícutin Volcano. The eruption started in 1943 and continued for nine years and twelve days to March 4, 1952, building a cinder cone to an elevation of 424 meters. The volcano edifice covers 25 square kilometers of land, but ash covers a much broader area. This photograph was taken on February 23, 1943. (Photo: R. Robles-Ramos; U.S. National Archives and the Smithsonian Institution)

buried by thick volcanic ash and lava flows from Parícutin and will never grow another ear of corn in our lifetime. In fringe areas around the volcano where the ash is thin and the lava did not reach, plants have invaded the devastated land, and agricultural activity has slowly recovered.

Before the eruption of Parícutin, volcanic deposits had provided nutrients for the rich soil that blessed the peaceful valley inhabited by the Tarascan Indians. Beginning in the sixteenth century, the Spanish colonialists forced the Indians to become farmers under their direction. Vulcan's soil supported corn and grasses for cattle and sheep, as well as kitchen gardens where residents raised vegetables, fruit, flowers, and herbs. The surrounding hills were covered with pine that provided resin and lumber as cash crops. Cattle and sheep grazed on the abundant grass that grew on the hills. Mary Lee Nolan, who did a classic study of the people affected by the Parícutin eruption, described the life in the valley: "In folk mythology the time before the volcano was a golden age when the region was beautiful and prosperous. Folk songs describe the flowers and singing birds, the rains of summer, and the green fields—a time when all was pleasure and happiness. . . . forests provided cover for the region's wild fauna, which primarily consisted of deer, rabbits, many species of birds, and wild bees valued for their honey. . . . The life of the region's people was not so idyllic as the folk songs suggest, but it was not a bad life" (Nolan, "Impact of Parícutin," p. 297).

After Parícutin erupted the villagers made valiant attempts to recover their agricultural life, but the 1942 corn crop, which had already been stored before the eruption, was used up in a few months. Livestock began to sicken and die. Wild fruit, bees, and deer began to disappear from the countryside. People averted famine only by accepting aid from relief agencies and from relatives living outside the area. Corn had been planted while the eruption was continuing in early 1943, but most of the plants were buried faster than they could grow, and the plants that survived burial were killed by fungi that entered plant tissues damaged by the ash.

In areas without lava flows, the thickness of the ash layer dictated survival of plants within Parícutin's area of influence. Where the ash was more than 1.5 meters deep, all living plants died. This zone of complete destruction extended well beyond Dionisio Pulido's farm. Where the ash was between 0.5 and 1 meter thick, trees and shrubs were heavily damaged. The ash stripped leaves directly from the trees or else formed a thin coating on broad leaves that restricted the access of carbon dioxide to the plant. Fruit trees as far away as 48 kilometers from Parícutin were affected. Fine ash prevented pollination, but it didn't matter, for there were no bees to pollinate the flowers—the ash had prevented their flight as it stuck to their bodies. Indirect effects of the volcanic ash included the elimination of a beneficial insect that preyed on sugarcane borers; as a result, 80 percent of the sugarcane crop was lost. Where the ash was greater than 14 centimeters thick, the blades of the oxen-pulled plows were not long enough to penetrate the underlying soil, and crops could not be planted (fig. 13-3).

Most preexisting shrubs and trees survived where the ash layer was less than a half-meter thick. And their chances of survival were significantly improved on steep slopes where the ash was washed away by rain. A benefit was the destruction of a fruit fly that had been damaging fruit. Where the ash was less than 3 centimeters thick, the roots of wheat and barley were able to reach the underlying volcanic soil and consequently provide high yields, but the thicker ash deposits remain barren to this day. The rugged lava flows contain small depressions that hold water, and by 1960, thirty species of plants, including pine trees, ferns, and mosses, were growing sparsely in moist areas on the lava. But lava cannot be plowed and therefore is unsuitable for farming.

Fig. 13-3. Where the ashfall from Parícutin was thin enough, oxen-drawn plows could mix the ash, like rich fertilizer, into the underlying soil. Where ash was too thick, nothing would grow and farms were abandoned. (Photo: Richard Barthelemy, University of Texas and the Smithsonian Institution)

The eruption ended in 1952. Most villages in the valley have been rebuilt elsewhere, and for several years the men supplemented their incomes with seasonal agricultural work in the United States in what was called the Bracero Program, that allowed Mexican migrant workers into the United States to harvest crops. Some villagers have moved back into the former eruption zone, but reestablishing themselves is difficult because the preexisting boundary markers were covered by ash fallout or lava flows and are not visible, and the situation frequently results in boundary disputes.

The Allure of Volcanoes

Farms

A frequently asked question is, "Why do people live on dangerous volcanoes?" The answer is easy. Volcanic soils are the richest on earth and people in need are willing to take high risks. In Indonesia, Italy, Japan, and the Philippines, farms cover the rich lower slopes of volcanoes. Where volcanoes are active, the farms can be repeatedly destroyed, but the farmers keep returning to them because they have no other land to cultivate.

Fig. 13-4. The pressure of population growth and demand for arable land have forced farmers near Legaspi City, Philippines, to farm the slopes of Mayon Volcano. On lower slopes rice and palms can be cultivated; the upper slopes are used for grazing and raising vegetables. The competition for land led to the death of seventy-five tomato farmers during an eruption in 1993. The farmers had cultivated small plots within restricted zones that are occasionally swept by pyroclastic flows and volcanic mudflows during eruptions. (Photo: Grant Heiken)

Occasional but common eruptions at Mayon Volcano in the Philippines illustrate the farmers' dilemma. Mayon is a composite volcano rising 2,900 meters above Albay Gulf near the Luzon city of Legazpi. Relatively small eruptions occur about every ten years, sending pyroclastic flows and lahars sweeping down the gullies that radiate from the summit crater. Beyond the gullies, downwind from the eruptions, a gritty scum of ash coats the land (fig. 13-4). The lower slopes of the volcano are covered with coconut plantations and rice fields, while tomatoes and other vegetables are raised on the intermediate slopes, and goats or sheep graze any open grassy areas. During the 1993 eruption of Mayon, seventy-five farmers were killed on their small tomato farms within Bonga Gully, one of the ravines of Mayon that had been declared off-limits by the government.

During the 1993 crisis, fifty thousand residents were evacuated from villages on Mayon's slopes, not entirely because of the destruction in Bonga Gully, but also in cautious anticipation of a possible larger eruption that occurs every one hundred to two hundred years at Mayon. However, the evacuations were not easily accomplished. Even if the land is at risk of destruction from an erupting volcano, removing a farmer from his land is like removing a hornet from its nest. Many residents of farming villages within 10 kilometers of Mayon's summit had to be forcibly removed to evacuation centers by the police. In spite of the danger and warnings, many people surreptitiously returned to their farms to tend their fields and farm animals during the day, because without their crops, they risked starvation.

Coffee Plantations

The next time you sip your coffee, reflect upon the problems of coffee farmers with plantations on the slopes of active volcanoes, and thank them for your coffee break. High-quality "hard bean" coffee is grown in most Central American countries between elevations of 1,400 and 1,700 meters (fig. 13-5). The best coffees are grown in volcanic soils in a tropical climate, where elevation moderates their growth. In addition to possible ruin from falling ash, the pyroclastic flows, lahars, lavas, and noxious gases that accompany many eruptions can destroy crops. In Central America, these conditions are commonly found on the slopes of young active volcanoes that have some of the best coffee-

Fig. 13-5. The massive, active volcanoes of southern Guatemala rise in the distance, east of Volcan Tecuamburro, the site of this coffee finca (plantation). Volcanic soils, tropical rains, and temperatures at elevations of 2,000 to 3,000 meters combine to produce some of the world's finest coffees. Without the volcanoes, there would be far fewer and poorer coffee fincas in this region. (Photo: Grant Heiken)

growing environments. During an eruption, such gases as sulfur dioxide and hydrogen sulfide can flow downhill and wilt the leaves of the coffee trees. Also, rainfall can extensively damage vegetation downwind from volcanoes that emit acidic gases as the gases dissolve in the rainwater to produce liquid acids.

Fertility and Rebirth

All life-forms on the earth's surface exist primarily by consent of nature's partnership—heat and light from the sun, elements from the rocks, and water. Moisture and gases from the atmosphere crumble the rocks and release elements in a soluble form that is easy for plants to incorporate for their growth—the first step in the food chain. It can be said that life is nurtured by nutrients derived from the rocks of earth by the process known as "weathering." It is easy to say, and in truth, "the earth is our mother."

Volcanic rocks are so rich in nutrients needed by plants that they can be considered as "hard" fertilizer. Lava flows as well as volcanic ash are equally rich, but it takes longer for nature to dissolve hard lava than fine-grained volcanic ash particles. A thin blanket of volcanic ash over a landscape can be considered to be a dusting of delayed-action fertilizer. Each glass shard in the ash is a time-release capsule, although it takes a bit too long after a shower of fresh volcanic ash to release the nutrients for immediate benefit to the farmer.

In warm, wet climates, rocks weather rather quickly, but in dry climates, warm or cold, weathering of rocks is extremely slow. Few places on earth illustrate the comparative effects of climate upon rocks in so small an area as the different islands of Hawaii. On the island of Hawaii, where most lavas have similar basaltic compositions, soil develops on congealed lava at sea level on the wet side (Hilo) within a hundred years, but on the dry side (Kona), even prehistoric lava flows several hundred years old show little signs of weathering.

The breakdown of Hawaiian basalt lava (and other rocks elsewhere) creates complex clay compounds that form the bulk of all soils, which anchor the root systems of plants and create a porous medium that retains moisture. Eventually the clay and mineral mixture becomes mixed with humic acids from decaying plant material, the last essential ingredient in the evolution of a rich soil. Thus, lava cools to make rock; thereafter, oxygen, water, carbon dioxide, and other molecules from the atmosphere interact with the minerals in the rocks (see chap. 2), breaking them down into soluble forms that roots can take up to nourish the plants that in turn feed the animals.

Rich volcanic soils have had profound effects on civilization, influencing population growth since mankind discovered agriculture because such soils are so fertile. One such place is the area around Naples, Italy, which is largely underlain by deposits of the Campanian Ignimbrite topped by yellow tuff (see chap. 5); these tuff deposits, composed mainly of glass shards, cover thousands of square kilometers from the sea into the foothills of the Apennines. The richness of the region was aptly described two thousand years ago by Pliny the Elder, the admiral of the Roman fleet who died during the Vesuvius eruption of A.D. 79. He wrote glowingly of the Campanian plain—saying that it "surpasses all the lands of the world." He noted that the land was kept in crop the whole year, being planted once with millet and twice with spelt, "and yet in spring, the fields having had an interval of rest produce a rose with a far sweeter scent than the garden rose, so far is the earth never tired of giving birth" (in Jashemski, "Pompeii and Mount Vesuvius," p. 592). Part of the rich growing area is on volcanic ash soils of Mount Vesuvius, but the vast growing area of the Naples region is largely over Campanian Ignimbrite. The fertile soil of Campania has been the source of economic prosperity for centuries, and the region is dotted with cities and large numbers of prosperous farms.

The rate at which volcanically devastated lands are reborn depends upon atmospheric and geologic factors. The most important atmospheric factors are annual temperature and rainfall, and one of the geologic factors is the chemical composition of the deposit (ash or lava). Basaltic material weathers to clay most easily; rhyolitic material weathers more slowly. The thickness of the volcanic deposit also plays a major role in recovery. If the deposit is thin enough, roots can find the preexisting soil layer, but if not, as someone once said, "Trying to grow plants in volcanic ash is like trying to grow them in styrofoam." Farmers have to wait several decades for the top of a thick newly deposited ash layer to develop into a rich soil.

Recovery from Vulcan's Devastation

Recovery of the devastated land at Mount St. Helens after the May 18, 1980, eruption (see chap. 1) began immediately but was only noticeable one year afterward. It started when plants sprouted from rootstocks in clear-cut or blowdown areas where ash deposits were thin and on hillslopes where vegetation was protected by snowdrifts at the time of the eruption. Many herbaceous plants are well adapted to sprouting from their below-ground parts after those above have been killed; they also grow rapidly in full sunlight, of which there was plenty in the blast zone. In the central part of the most devastated area, mudflows from steep slopes had brought mud mixed with seeds and preeruption soils down onto flat valley floors to form islands of vines and assorted herbaceous plants sitting on top of 200 meters of sterile ash. One of us (RVF) spent the summers of 1982–1987 doing research in the Mount St. Helens blast deposit and watching the gradual encroachment of natural plants, mainly fireweed (*Epilobium augustifolium*) and ferns, which started on the edges of the devastation area on very thin ash layers and encroached toward the regions of

Fig. 13-6. The slopes of Mount St. Helens are barren ground with a few isolated plants. This drawing of the prolific Pacific Northwest fireweed (*Epilobium augustifolium*) plant is a signal of rebirth after the destruction of the plants and animals by the eruption.

thicker ash (fig. 13-6). Replanting of pine trees was successful only where the holes could be dug into the underlying soil. Today, seventeen years later, natural and assisted growth on Mount St. Helens' deposits is gradually softening the bleak, desert-like appearance of the once verdant, evergreen-covered land.

Even isolated new islands formed by eruptions in the open ocean are rapidly populated by plants. Surtsey, a new island formed by eruptions off Iceland's southern coast in 1963, now has eight species of blue-green algae, three species of moss, and vascular plants growing near fumaroles, where there is condensed water, but colonization is slow because of the frigid environment.

In tropical environments, recovery is much faster and the vegetation is more luxuriant. One of the earliest studies of the recolonization of a volcanically devastated landscape was on the tropical island of Krakatau, which was nearly obliterated during its 1883 eruption. The parts of the island that survived were left barren. In 1919, Krakatau was visited by Dr. W. M. Docters van Leeuwen, professor of botany at the Medical Facility of Batavia and director of the Botanical Gardens at Buitenzorg. He reported that the island was not yet covered by a continuous forest—there were only small pockets of trees surrounded by grass-covered slopes. But ten years later, the trees had spread, merging into a continuous forest that killed off the grasses in their shadow. Grasses survived only in small patches on the driest of slopes. The present-day jungle is now so thick that it prohibits detailed geological exploration of the 1883 deposits.

On lava flows, soils develop more slowly than on volcanic ash. By definition, volcanic soils derived from volcanic ash are *andosols*, a Japanese word (*an* = "dark"; *so* = "soil"). They are good, lightweight insulators that protect plant roots from rapid temperature changes, have an open porous quality, contain ample organic matter relative to the soil's weight, have considerable clay, can transfer and exchange elements in solution quite easily (from water to roots), have an ability to retain phosphorous, and have a high water content. In volcanic soils from Indonesia to Japan, root crops such as radishes, sweet potatoes, and carrots do well, and upland rice can be grown because of great water retention of the andosols. The andosols of the Pengalengan Highlands support the best tea plantations in Java. In Hawaii, soils formed on volcanic ash are excellent for all crops except pineapple, which thrives mainly on soils derived from lava flows. Sugarcane, papaya, and coffee grow well on the warm, moist ash soils; the drier soils support the grasslands used as cattle ranges.

The fertile farmlands of New Zealand are on volcanic soils of different ages—the soil in the Waikato area is four thousand years old and that in the Bay of Plenty area is forty thousand years old. Combined with abundant rainfall, warm summers, and mild winters, these regions produce abundant crops such as the kiwifruit. The volcanic ashes, altered mostly to clays, are well drained yet moisture-retentive, and are easily tilled. The deep volcanic loams are particularly good for pasture grasses, horticultural crops, and maize.

The central part of New Zealand's North Island is underlain by young volcanic terrain called pumice lands by the local people. One eruption of extraordinary size occurred at Taupo and covered much of the North Island with pumice-rich pyroclastic flows and pumice fallout. But when settlers first moved into the pumice lands, mainly to graze sheep, the cobalt, selenium, and copper deficiencies in the young immature pumice soils, consequently in the plants, caused "bush sickness" and death of the livestock. In the mid-1930s, the government and the logging companies established large plantations of pines (*Pinus radiata*) to make use of the pumice lands. Pumice soil facilitates good drainage, yet retains enough moisture for the rapid growth of pines. Moreover, the deep-rooted pines reach through the pumice deposits into the older soils for nutrients. These pine forests are now being harvested sixty years later for use in export lumber and paper pulp. Cobalt-bearing phosphate fertilizer is now used to counteract the mineral deficiency and solve the problem of bush sickness on grazing lands.

References

Blong, R. *Volcanic Hazards: a Sourcebook on the Effects of Eruptions*. Sydney, Australia: Academic Press, 1984.

Jashemski, Wilhelmina F. "Evidence of flora and fauna in the gardens and cultivated land destroyed by Vesuvius in A.D. 79." In *Volcanism and Fossil Biotas*, edited by G. Martin and A. Rice, 113–25. Geological Society of America Special Paper 244. Boulder Colo., 1990.

―――. "Pompeii and Mount Vesuvius, A.D. 79." In *Volcanic Activity and Human Ecology*, edited by P. D. Sheets and D. K. Grayson, 587–622. New York: Academic Press, 1979.

Luhr, J. F., and T. Simkin. *Parícutin: The Volcano Born in a Mexican Cornfield*. Phoenix, Ariz.: Geoscience Press, 1993.

Molloy, L. *Soils in the New Zealand Landscape: The Living Mantle*. Canterbury: New Zealand Society of Soil Science, 1993.

Nolan, M. L. "Impact of Parícutin on five communities." In *Volcanic Activity and Human Ecology*, edited by P. D. Sheets and D. K. Grayson, 293–338. New York: Academic Press, 1979.

Sheets, P. D., and D. K. Grayson. *Volcanic Activity and Human Ecology*. New York: Academic Press, 1979.

Simkin, T., and R. S. Fiske. *Krakatau 1883: The Volcanic Eruption and Its Effects*. Washington, D.C.: Smithsonian Institution Press, 1983.

Ugolini, F. C., and R. J. Zasoski. "Soils derived from tephra." In *Volcanic Activity and Human Ecology*, edited by P. D. Sheets and D. K. Grayson, 83–124. New York: Academic Press, 1979.

Wilcox, R. E. *Some Effects of Recent Volcanic Ash Falls with Especial Reference to Alaska*. U.S. Geological Survey Bulletin 1028-N. Washington, D.C. 1965, pp. 409–76.

Volcanic Rocks: Guardians of History

Frontispiece. In A.D. 79, ash from Mount Vesuvius flooded the land around it and variously destroyed, buried, and preserved many villas and villages. This photograph shows an excavation into the A.D. 79 deposits 6 kilometers from the crater of Vesuvius. It reveals the way of Roman life and has preserved and guarded the information for nearly two thousand years. (Photo: Richard V. Fisher)

Volcanic Ash: The Nearly Perfect Preservative

Paleontologists and anthropologists spend hours, days, and years scouring the ground for fossils and artifacts, often in inhospitable lands under great hardship. But the excitement of discovering evidence of early man or making other finds is a seductive reward—one that would be lost in many cases were it not for volcanic eruptions. Burial of a farm, village, or field by lava flows, pyroclastic flows, and lahars causes great damage and generally destroys everything on the ground, but thick falls of volcanic ash are a nearly perfect preservative. Early human history has been preserved and guarded by ash layers for millions of years, and archaeological excavations of such sites have revealed the history of various peoples and their ways of life.

A constant and rapid fall of volcanic ash from a large eruption is ideal for preserving what exists upon the land at the time of deposition. It can bury them intact without moving them and without completely crushing or burning them. No other natural catastrophe is capable of simultaneous burial and intact preservation. Dust storms that coat surfaces with fine sediment are the nearest analog to fallout ash, but the amount of falling dust delivered by a storm cannot create a deposit thick enough to bury all objects, as can a voluminous fall of ash.

Why are these usually dull, muddy rocks so rich in fossils? During an explosive volcanic eruption, the land is covered with a blanket of volcanic ash. If it is thick enough, it will suffocate much of the animal life outright or eliminate it indirectly by killing the plants needed by the vegetarians, who then starve and in turn deprive the carnivores of prey. Heavy falls of ash also affect water runoff, which drastically alters the landscape—instead of soaking into the ground, rain drains away quickly. The runoff is laden with volcanic ash that covers a flood plain farther down in the valley with thick muds that quickly bury plants and animals—exactly the conditions needed to fossilize the flora and fauna for the geologic record. Water that seeps through the deposited ash reacts with glass shards and dissolves or replaces many of them with other elements to create either void space or derivative minerals, such as clay minerals. Excess silica dissolved in the water often replaces soft bone materials molecule by molecule, creating an almost indestructible fossil (like petrified wood) that lasts for millions of years. Leaves commonly are preserved only as imprints. Volcanic-ash deposits not only cause the widespread death, but then also provide the substance that preserves the bones of the buried animals.

Human Origins in the
Ethiopian Rift

Over millions of years, numerous eruptions in or near the Ethiopian rift have produced volcanic ash deposits beneath which are preserved the remains of early man and other animals. Once a verdant place, it is now a hostile and thorny land that hides bones and makes life difficult for those who seek them (fig. 14-1).

The East African Rift is a 4,100-kilometer-long series of deep valleys, lakes, and rivers that divides the continent—from Mozambique to the Red Sea (itself a flooded rift). The rift was formed as extensional forces slowly pulled apart the African continental plate. The process has continued for millions of years and is still occurring at the geologically rapid average rate of 1–2 centimeters per year. As the crust separated, low-lying valleys hundreds of kilometers long with highlands on their sides were created, and magmas rose to leave lines of volcanoes at the surface. The oldest volcanoes were in the center of the rift, but they were pulled apart, and now the different parts of many once whole volcanoes occur on opposite sides of the rift in the marginal highlands. Eruptions of

Fig. 14-1. The East African Rift, pulled apart by colossal forces, has experienced volcanic eruptions off and on for millions of years. The land consists of layers of volcanic ash, ashy sediments, and lavas. The earliest hominids, ancestors of modern human beings, thrived in this lowland region but were occasionally buried by volcanic ash or floods from erupting volcanoes. Volcanic ash beds and ash mixed into sediments have preserved fossils in the lowlands of the region and also are used to date the rift fossils by radiometric methods. The oldest known hominid (*Ardipithecus ramidus*) lived here 4.4 million years ago. (Photo: Grant Heiken)

the rift volcanoes have filled the valley basins with ash, burying and fossilizing the numerous animals that once roamed the valleys.

Volcanoes of all types occur in the dynamic rift setting. The most common ones are cinder cones and hydrovolcanic tuff rings (in lakes) with their associated widespread basaltic lava flows. Less frequent, but more important to the fossil-hunting paleontologists are the large calderas from which silica-rich magmas generated the large-volume explosive eruptions that layered volcanic ash over animals and plants (fig. 14-2).

The Ethiopian rift widens and divides to form the Horn of Africa—the northern segment continues into the Red Sea, while the east-trending segment

Fig. 14-2. The bleak badlands topography of the East African Rift can be inhospitable but is a natural museum that holds rich treasures of human development. Here, geologists examine and sample volcanic ash deposits interbedded with fossil-rich sediments in the Afar Rift of Ethiopia. (Photo: Grant Heiken)

extends into Djibouti, a desert nation with no resources other than remarkably arid terrain, hot springs, young and rocky fault scarps, and volcanoes. Along the rift is a series of sinking basins that have preserved the fossil record and at times are faulted upward along the rift's margins. The faults slice into and expose the deposits sought by paleontologists and geologists.

Most of the discoveries of hominid remains have been found along the East African Rift in the last seventy years. Decade after decade of field research has pushed back the time scale for the australopithecine species. The best-known finds in the last twenty years have been *Australopithecus afarensis* (Lucy), discovered at Hadar in 1974, and *Ardipithecus ramidus*, discovered at Aramis in 1994; both finds were within the northern Ethiopian rift. Lucy has been dated at 3.4 and between 3.89 and 3.86 million years B.P. (before the present) and *A. ramidus* at 4.4 million years B.P., by radiometric dating of the volcanic ash beds that lie above and below the fossils. *A. ramidus* is the most apelike of the hominid fossils found so far and may be close to a common ancestor for African apes and humans.

Many of the large composite volcanoes that now mark the rift center did not exist during the time of *A. ramidus*. Instead there were many smaller volcanoes, ranging from cinder cones with lava flows to hydrovolcanic tuff rings along the shores of ancient lakes. At that time, there was a wooded habitat within the rift that was far more hospitable than the present-day harsh land. Occasional ashfalls from distant volcanoes dusted the area, but these were mere annoyances. It was the catastrophic fall of thick ash layers that disrupted the land and buried the animals, plants, and hominids for posterity.

One difficulty in determining the ages of fossils is that by themselves, fossils do not tell absolute time—only relative age can be resolved by noting their positions in the rocks. For example, fossils in lower layers are obviously older than those in the upper layers. It is necessary to use other methods to determine their absolute ages, and ash beds are a prime site for radiometric dating of fossils, for they contain feldspars that have radiogenic atoms. In the rifts, the fossil-bearing sedimentary rocks are interbedded with volcanic ash beds and lavas. Since an entire volcanic ash layer is deposited almost simultaneously, it is the same age everywhere it is found. And, since each ash bed is often unique chemically and mineralogically, they can be identified across thousands of square kilometers through a combination of field observations and laboratory analyses of specimens. Ash beds with potassium-bearing minerals, such as some feldspars, can be dated by the potassium-argon method. The radioactive decay of the potassium forms argon, which can be measured in the laboratory. The half-life of potassium can be used to accurately measure the moment when the potassium-bearing minerals in the volcanic rocks were formed.

A better understanding of human evolution has come from discoveries made in this century, most of which have been along the East African Rift. The unique combination of rifting, of more or less continuous volcanism, of thick layers of ashfall tuff, and of rift sedimentation has provided not only a fertile environment for life, but the combination of geological factors needed to preserve the record.

Preserving Historic Cultures

The Circumvesuviana railroad from Naples to the excavations at Pompeii passes through such densely populated towns on the southern slopes of Vesuvius as Portici, Ercolano (Herculaneum), Torre del Greco, and Torre Annunziata. Nearly a million people live in this unique rural setting, all within range of pyroclastic flows that could again descend from the summit of Vesuvius as they did in A.D. 79 when they buried Pompeii, Herculaneum, and many villas on and around Vesuvius (fig. 14-3). Although these towns are now mostly populated by workers from the industrial plants located on the outskirts of Naples, the local people have roots in the rich volcanic soil.

For centuries, people have been drawn to the region by its fertile soil and proximity to the Bay of Naples. The train route over the plains below Vesuvius crosses intensely tilled land—every square meter of this rich soil is used. For example, the small vineyards have grapes and spring beans on the trellises and

Fig. 14-3. The excavated ruins of Herculaneum (lower half of photograph) buried beneath a pyroclastic flow on the west side of Mount Vesuvius in A.D. 79. A modern city, which was damaged by the southern Italian earthquake of 1980, was built upon the once barren pyroclastic plain above Herculaneum (upper part of photograph). The two cities, separated in time by two thousand years, look very much alike. (Photo: Richard V. Fisher)

Fig. 14-4. Vineyard on the slopes of Vesuvius today. Agricultural techniques and the growth mixture of grapevines, vegetables, flowers, and nut tree species have not changed much in the nearly two thousand years since A.D. 79. (Photo: Grant Heiken)

fava beans, cauliflower, and onions between the trellis rows, and the vineyard margin is rimmed with orange and lemon trees, herbs, and flowers. The pale golden wine Lacrimae Christi ("tears of Christ") is made from grapes grown on the southern slope of Vesuvius. The region is also noted for tomato growing (fig. 14-4).

Judging by prolific writings and paintings, agricultural life around Vesuvius has remained more or less the same since Roman times, as the poet Martial enthusiastically described it: "This is Vesuvius, not long ago green with vine leaves where the golden grapes pressed the wet casks. This is the mountain more beloved by Bacchus than the hills of his native Neisse. On this mountain just a moment ago the satyrs performed their dances. This (Pompeii) was Venus' seat which she preferred to Sparta; this other place (Herculaneum) was famous for the name of Hercules" (quoted in Paliotti, *Vesuvius*, p. 37).

Not only have the rural lifestyle and vistas remained the same over the last two thousand years, but so have the methods of viticulture (fig. 14-5), as was confirmed in the mid-1960s by Wilhelmina Jashemski, from the University of Maryland, who excavated a vineyard at Pompeii that had been buried by pumice falls during the A.D. 79 eruption of Vesuvius. In 1966, she began to excavate the Foro Boario (Cattle Market—an early name used by excavators), a

Fig. 14-5. Large terra-cotta wine bottles in a villa excavated on the southern plain of Mount Vesuvius. They were buried in A.D. 79 during destruction of Pompeii. (Photo: Richard V. Fisher)

villa, where she found hundreds of vine root cavities and adjacent grape-stake cavities. Also uncovered were paths dividing the vineyard and cavities on each side of the paths once occupied by posts that supported an arbored passageway. In 1972, in the villa's garden House of the Ship Europa, Jashemski's team found two plots with distinct furrows in a lower garden and many single cavities, evenly spaced, with distinct depressions for holding water on each side. The furrows may have been for vegetables, but the roots and stems had been too small to be preserved. Root cavities for small fruit trees and vines occurred 4.5 Roman feet apart. Pots were embedded in the soil at intervals along the garden walls. Also uncovered were carbonized fruits and vegetables—almonds, filberts, grapes, grape seeds, broad beans—mixed plantings like those found in today's gardens (fig. 14-6).

Further work in the Garden of Hercules revealed embedded pots, roots of citron or lemon trees, carbonized cherries, olive-tree roots, and pollen. Soil contours divided the Roman garden into many beds, and the archaeologists believe that this may have been a large commercial flower garden, similar to those found today between Pompeii and Naples. It was likely that the flowers from Pompeii were used for making garlands and perfume and graced the interior of wealthy villas, as well as the tables of the lowly peasants who did the work.

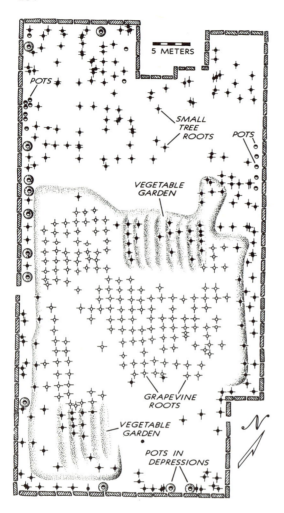

Fig. 14-6. An excavated vineyard at Pompeii, buried in A.D. 79 indicates that viticulture methods have not changed much in two thousand years. This diagram of the House of the Ship Europa shows two plots of ground with distinct furrows and many evenly spaced cavities with distinct depressions for holding water on each side. Pots were embedded in the soil along the garden walls. There were mixed plantings of many fruits and vegetables, like those those found in today's gardens, passed down from farmer to farmer over the centuries. (Adapted from Jashemski, 1979)

Fig. 14-7. Wine bottles with logos of volcanoes and their products are among the collectible souvenirs at vineyards in volcanic regions. These are wines produced on the Island of Thera (Santorini), in the southern Aegean, Greece. (Photo: Grant Heiken)

Another snapshot of an even more distant past, Bronze Age Greece, has been preserved by volcanic ash from a large caldera-forming eruption in 1628 B.C. (see chap. 10). At Akrotiri, on the island of Thera, excavations have revealed the agricultural practices of 3,600 years ago. An archaeological team led by Christos Doumas, from the University of Athens, found that ancient Therans practiced animal husbandry and were avid meat-eaters. Animal bones found at an excavation site indicate that the Therans preferred sheep and goats in their diet, pigs second, and cattle third. The lack of the bones of cattle suggests that they were raised as beasts of burden and perhaps for milk. Game animals, mostly boars, were also part of the Theran diet, and the remains of their bones prove that these ancient Greeks hunted.

At Akrotiri, storage jars were uncovered that contained the carbonized remains of legumes (lentils and split peas or favas), bread made from barley flour, imported olives, sesame seeds, saffron, almonds, and pistachios. Honey was gathered for use as a sweetener. Wine was made for local use, but it was also an item for export; today, wines are still made on Thera, but little of the harvest ever leaves the island—many of the powerful (15 percent alcohol) wines produced today are descendants of the island's volcanic heritage, with labels such as "Caldera," "Volcano," and "Lava" (fig. 14-7). It is also highly probable that the people brewed beer, for it was a popular Egyptian beverage

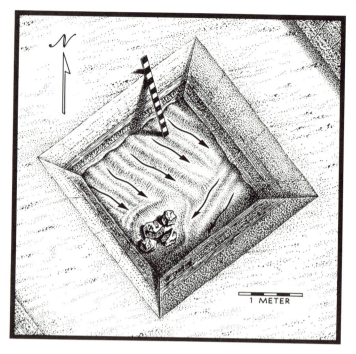

Fig. 14-8. In Zapotitán Valley in west-central El Salvador, research was conducted to determine recovery rates of soil and vegetation from a heavy fall of ash. This diagram is of a test pit through a sheet of ash that buried the ground in the sixth century A.D. The test pit, 20 meters southwest of a buried farmhouse, reveals furrows and ridges. Ridging is done to retain rainfall moisture early in the rainy season. Casts of corn plants 5 to 10 centimeters high were found on the tops of ridges. Their height suggests that for El Salvador, it was early in the season and the eruption that buried the field and farm most likely occurred in May. Rocks in the lower corner are the broken fragments of a large, hot cinder block that fell into the cornfield early in the eruption. (After Sheets, 1979, figure 17.12)

at the time. As more of the thick deposits of volcanic ash are removed, teams of archaeologists and agronomists are finding out more about farming in Minoan times. The research proves that ancient habits don't die. As in the case of Pompeii, little has changed in farming techniques on the arid islands of the southern Aegean Sea.

In the New World's Mayan civilization, farming techniques also have been preserved in various places beneath volcanic ash. In the mid-1970s, a team led by Payson Sheets, from the University of Colorado, began a methodical search for Mayan villages buried by volcanic ash in the Zapotitán Valley of west-central El Salvador. The valley has felt the effects of several volcanoes through Mayan and Spanish colonial times.

A Mayan farm at Ceren was unearthed after removing four meters of lava bombs, pumice, and ash fallout that had been dumped on the farm and fields by an eruption of Laguna Caldera in the sixth century. Unlike Europeans, who lived in central villages for protection during feudal times, the Mayan farmers lived in isolated farmhouses located near their fields. The multi-room farmhouse was constructed of fired-clay floors and columns, wattle-and-daub walls, and a palm-thatch roof. People had been trapped and killed in the house by the hot ash. Ceramic jugs that were found were full of beans, and perhaps had contained beer made from maize. A spindle whorl indicated that garments had been woven from cotton. Plant casts of what may have been maize were found on low ridges spaced at half-meter intervals. The fossil leaves are long, narrow, and coarse, like those of young maize. If the field had not been irrigated before it was buried, then the height of this plant indicates that the eruption occurred in May (fig. 14-8).

THERE is no doubt about the fertile gift of Vulcan, but volcanic eruptions are more often viewed as destructive events because in the short term they destroy human lives and habitations or bury rich, life-giving soils. Yet, burial preserves evidence of ancient lifestyles for future generations to study and learn about their heritage. And in the long-term, volcanic eruptions provide rich agricultural lands and, as shown elsewhere in this book, many other benefits. We have summed up the beneficial aspects of volcanism, and in the last chapter, we will discuss ways to save lives during a volcanic eruption.

References

Blong, R. *Volcanic Hazards: A Sourcebook on the Effects of Eruptions.* Sydney, Australia: Academic Press, 1984.

Jashemski, W. F. "Pompeii and Mount Vesuvius, A.D. 79." In *Volcanic Activity and Human Ecology*, edited by P. D. Sheets and D. K. Grayson, 587–622. New York: Academic Press, 1979.

Paliotti, V. *Vesuvius: A Fiery History*. Napoli: Cura E. Turismo, 1981.

Sheets, P. D., and D. K. Grayson, eds. *Volcanic Activity and Human Ecology*. New York: Academic Press, 1979.

PART IV

Living near Volcanoes

Mitigation and Survival

Frontispiece. Volcanologist at work collecting gas samples at St. Augustine Volcano, Alaska. Composition of the gases may give clues to the rate at which magma is approaching the surface. (Photo: Harry Glicken)

Fig. 15-1 (*Top*). In 1984, two years after the El Chichón, Mexico, eruption, local men ride across the former site of Volcan Chichón. Bushes in background are beginning to invade the devastated land. (Photo: Richard V. Fisher)

Fig. 15-2 (*Bottom*). Site of the village of Volcan Chichón, which was reduced to a dune field by a pyroclastic surge. The pole at left is actually a rooted, broken, and sandblasted tree. The pyroclastic surge traveled down the valley (away from the reader) for another five kilometers. The village was located in the approximate center of the photograph along a river bank. (Photo: Richard V. Fisher)

W̲e watched the figures with straw hats shading their dark faces as they silently rode back and forth on mules across a barren plain at the foot of El Chichón Volcano (fig. 15-1). The plain had been the site of their village, and the people appeared to be searching for signs of loved ones and possessions. Perhaps something had been uncovered by the wind and the rain during the two years since the eruption in 1982 that had come in the night and taken everything from them (fig. 15-2).

El Chichón was once a heavily vegetated hill in an obscure region in the state of Chiapas, southern Mexico, and there had been several villages at the foot of the volcano before the eruption. With an elevation of 1,350 meters, El Chichón stood about 500 meters above the surrounding countryside and had a crater partly filled with water and containing a 400-meter-high dome. As indicated by radiocarbon dating, the volcano had last erupted about six hundred years ago, and further radiocarbon dating after the 1982 eruption revealed that El Chichón erupts about every six hundred years. But the villagers did not regard the volcano as dangerous, despite the hot springs and steaming fumaroles that had been emanating from it for as long as they could remember. The abundant vegetation on the volcano's slopes showed that no large eruptions had occurred in recent centuries. The ground was fertile and life was simple. In that part of the world, land provides the only livelihood, and continuous hard work is required to produce a healthy crop. It is unthinkable for villagers to leave their crops to nature's whims, especially when no one can guarantee that a life-threatening situation will actually materialize. The land is the life's blood of a poor farmer, and he and his family cannot leave it and survive without government subsidies, large shelters, or imported food; if the government cannot provide assistance, the people are bound to the land.

In the autumn of 1981, the people began to feel earth tremors, and their dogs became restless. From time to time, small earthquakes rattled dishes, and breezes occasionally wafted the rotten-egg smell of hydrogen sulfide gas. The unusual events were a topic of chatter in the tiny village plazas, where people bought local produce and gossiped in the hot, lazy afternoons. Their alarm would have been much greater had they known the shudderings and smells were the prelude to an eruption. Two Comisión Federal de Electricidad (CFE) geologists doing fieldwork at the volcano between November 1980 and April 1981 felt earthquakes and were frightened by loud noises. In their report, they suggested the possibility that the volcano was reawakening. Their report was filed, but because the CFE is not concerned with volcano hazards, no action was taken.

After the eruption, study of the seismographs showed that the tremors had increased in early 1982 and that the earthquakes had risen from depths of 5 kilometers to 2 kilometers. On March 29, at 5:15 A.M. local time, the pre-dawn sky was shattered by an enormous roar and nearly continuous earthquakes. A powerful ash plume rose thousands of meters above the volcano's summit. The roaring Plinian eruption continued for six hours, and lightning crackled through the ash cloud, assaulting the eardrums of the villagers and intensifying their fears. Falling rocks punctured rooftops, and ash collected in deep layers everywhere, collapsing houses. Hot gases caused a jetlike rise of the eruption cloud into the atmosphere, where the ash cloud was driven by high-altitude winds to scatter ash northeastward across southern Mexico. The eruption reamed out the center of El Chichón, transforming the thickly vegetated domelike hill into a barren, kilometer-wide crater 300 meters deep with near vertical inner walls.

The new El Chichón crater continued to widen by minor steam explosions, until a second and a third eruption pierced the sky on April 4. The eruption columns were too heavy to rise into the atmosphere and therefore collapsed downward. Like a giant hand that opens and lets fall its contents, the collapsing column dropped tons of ash, lapilli, and blocks the size of cobbles. The mass of material had great momentum and therefore could not stop as it fell onto the volcano's summit and upper slopes. The pyroclastic flows and surges plummeted down the volcano's slopes and through the valleys surrounding the volcano, furiously ripping apart and carrying everything away; what remained was buried in ash. The surge activity of April 4 was the devil's own curse, for on that day, more than two thousand people died and at least nine villages within an 8-kilometer radius were obliterated (fig. 15-3).

The residents who had chosen to remain in the villages found some cover from the ash fallout but had no place to hide from the pyroclastic flows or the surges. Homes with single-wall wood construction or corrugated-iron sides were no match for the fierce hurricane-like pyroclastic currents, which spread across the face of the volcano as far as eight kilometers from the crater. One set of surges devastated an area of about 100 square kilometers and another devastated an area of 40 square kilometers.

When the volcano began to awaken, its catastrophic potential was unknown because volcanologists had not yet determined the recurrence rate of its eruptions nor did they know the nature of its past activity. It was this lack of information about the volcano that diminished the urgency to evacuate the residents from the several nearby villages. Because the residents were not aware of the risk, most of them stayed. This disaster and its circumstances profoundly demonstrate that the earth's active or potentially active volcanoes need to be systematically studied in all nations.

Fig. 15-3. *Top*: The crater of El Chichón was once a verdant hill, which had been a dome that had filled an older crater. The eruption cored out the dome. The rim of the former volcano now encloses the crater lake of El Chichón. (Photo: Courtesy of Harry Glicken). *Bottom*: Fumaroles surround El Chichón's sulfur-charged crater lake with yellow sulfur beaches. A small, hot magma body sits beneath the surface heating the lake waters. (Photo: Richard V. Fisher)

Assessing the Hazards: The
Pulse of Volcanoes

Volcanoes of the world are now being indexed for their hazards and risks, an endeavor encouraged by the United Nations, which declared the years 1990– 2000 to be the International Decade for Natural Disaster Reduction (IDNDR). The purpose is to foster research on the causes and effects of various kinds of disasters. One theme is the Decade of the Volcano, which has the intent of promoting greater knowledge of volcano behavior, expecially why and how often a volcano might erupt. The fifteen volcanoes targeted for concentrated research during the decade are among the most famous and potentially the deadliest in the world: Mauna Loa and Mount Rainier (United States), Colima (Mexico), Santa Maria (Guatemala), Galeras (Colombia), Nyiragongo (Zaire), Mounts Etna and Vesuvius (Italy), Ulawun (Papua, New Guinea), Taal (Philippines), Merapi (Indonesia), Mounts Sakurajima and Unzen (Japan), Thera [Santorini] (Greece), and Teide (Canary Islands, Spain).

Volcano studies are a wise investment of public money because they can save lives. An example of renewed modern research on an active volcano is Mount Rainier, a Decade Volcano that was constructed over a period of about a million years. In the last 10,000 years it has erupted many times, the most active periods being 6,500–4,500 years ago and 2,500–2,000 years ago, as determined from radiometric dates of its layered deposits. Because volcanoes are huge masses of rock commonly complexed with lava flows and pyroclastic deposits, how can geologists find ways to study them and predict their hazards? One way is to geologically "take the pulse" of a volcano.

An aim of studying a volcano during its inactive years is to find its baseline of "normal" behavior. With the norm established, any changes alert volcanologists to possible activity. Particularly when there is a large population near a potentially active volcano, there is an urgent need to watch its behavior. In such populated areas as the Seattle-Tacoma region, even a small- to medium-sized eruption or a medium to large lahar could result in a great loss of life and tax the resources of the region (see chap. 6).

Mount St. Helens is a prime example of how volcanologists determine a volcano's history. In the early and mid-1970s, Rocky Crandell and Don Mullineaux, both of the USGS, climbed most of the ridges and valleys around Mount St. Helens, mapping, describing, and measuring the thickness and distribution of the volcanic rock strata from which Mount St. Helens volcano is built. Their purpose was to reconstruct the eruptive history of the volcano, to determine the kinds of hazards expected from the volcano, and to collect information for hazards mitigation should an eruption occur.

By 1978, Crandell and Mullineaux had finished their research on Mount St. Helens and had published their findings. They found that most of the volcano was only a thousand years old and that it had been constructed on top of

Table 15-1. Eruptions of Mount St. Helens A.D. 1400–1980

Date*	Type of Activity	Date*	Type of Activity
1980	modern eruption	1600	no activity
1850	dome and explosive eruptions	1550	no activity
1800	explosive eruptions, mudflows	1500	ash fallout and pyroclastic flows
1750	no activity	1450	no activity
1700	no activity	1400	pyroclastic flows
1650	pyroclastic flows, mudflows, dome eruption		

Source: Information from Crandell and Mullineaux (1978).
* Approximate.

an older volcanic center that first erupted about 36,000 years ago. During the last 600 years, Mount St. Helens had been more active and more explosive than any other volcano in the coterminous United States. Eruptions had repeatedly formed lava domes, had explosively expelled hot pyroclastic flows and large volumes of pumice, and had caused many mudflows. Crandell and Mullineaux determined the eruption frequency—the pulse of the volcano—and, with that knowledge in 1978, forecast that it could erupt before the year 2000 (table 15-1):

> The present dormant state of Mount St. Helens began in 1856, and no way is now known of determining when the volcano will erupt again. Mount St. Helens' behavior pattern during the last 4,500 years has been one of spasmodic periods of activity, separated by five or six dormant intervals of a little more than 2 to about 5 centuries duration. In addition, 12 dormant periods 1 to 2 centuries in length have been identified, and many intervals of a few years or a few decades surely occurred during prolonged periods of intermittent eruptive activity. The volcano's behavior pattern suggests that the current quiet interval will not last as long as a thousand years; instead, an eruption is more likely to occur within the next hundred years, and perhaps even before the end of this century. (Crandell and Mullineaux, *Potential Hazards*, pp. 24–25.)

People that climb a fuming volcano take a statistical risk—usually a small one, but still a risk. The average person cannot readily assess such risks, but volcanologists can help by taking the pulse of volcanoes. The relative ages of layers within a sequence of rocks can be readily determined by simple field observations—the bottom layer of rock is older than the one above it, and so on, up through the sequence. And then, the absolute ages of the rocks are determined by measuring the amount of radioactive substance left in the mineral or rock against radioactive substances of known half-life. If the half-life of a radioactive element is one million years, and half of it has spontaneously broken down, then the rock is a million years old. By determining the dates of active volcanism indicated by the deposits, it is possible to tell how many years have elapsed between eruptive periods (see chap. 14).

Long-quiet volcanoes that become active usually show some signs upon reawakening and so give people time to prepare for an eruption. But active volcanoes with fumaroles and occasional explosions are a different story, for short-term forecasting is less certain unless the precursor signs are well known through long-term studies. One thing is certain about volcanoes, however: eruption times cannot be precisely predicted because they are chaotic, non-linear systems, meaning that a particular sign of awakening may lead to different consequences at different times. Awakening activity does not necessarily mean that there will be an eruption, and a quiet volcano is capable of erupting with little warning.

The location of a volcano and the composition of magma that feeds it determine the frequency and types of eruptions. Oceanic volcanoes fed by hot spots in the crust commonly have frequent nonexplosive eruptions of slow-moving basaltic lava that often begin with spectacular lava fountains, for example those on the island of Hawaii and in Iceland. The Laki eruption of 1783 poured out 14 cubic kilometers of lava from a 25-kilometer-long fissure, and the chemicals pumped into the air by that eruption created a European environmental disaster (see chap. 9).

What Is a Disaster?

Everyone has their personal view of how to define a "disaster," but there is no generally accepted definition. As stated by O'Keefe and Westgate: "[A] 'disaster' refers to a 'manifestation of an interaction between extreme physical or natural phenomena and a vulnerable human group' that results in 'general disruption and destruction, loss of life and livelihood and injury' " (in Tilling, *Volcanic Hazards*, fig. 1.1, p. 3).

The terms *disaster*, *hazard*, and *risk* are difficult to define, but it is necessary owing to the need to prioritize limited economic resources for distribution.

A volcanic *hazard* is a destructive natural process that has occurred previously at a particular volcano; therefore, there is the probability that it could occur again during a future eruption. If the hazard has occurred several times at the same volcano, its recurrence is very likely. For example, if lahars occurred once during the last ten eruptions, there is a 1-in-10 probability that a lahar will occur in the next eruption.

Risk is a measure of the potential loss from a hazard, and losses can include people's lives, property, livestock, and the productive capacity of the area, therefore the magnitude of the risk increases as population increases near a volcano. More precisely, risk can be calculated from the following formula: risk = (value) × (vulnerability) × (hazard). *Value* is the number of threatened lives, property, civil works, and productive capacity; *vulnerability* is a percentage of the value likely to be lost during an eruption, and *hazard* is a probability (1 in 10, etc.). Villages in Indonesia close to active volcanoes have a high risk rating because eruptions occur frequently, villages are quite vulnerable, and fields and buildings, along with their economic value, can be destroyed. Precise numbers can be assigned to risks within the formula above. Probability cannot be used to accurately predict what will happen, but forecasts can be used to consider the necessity of an evacuation. Whether or not there is an evacuation depends upon the type of expected volcanic activity. If a volcano has a history of highly explosive eruptions and the probability that an eruption will occur is calculated to be a 1-in-10 chance, it would be prudent to evacuate. But if the probability is calculated at 1 in 10,000, the recommendation to evacuate would be less urgent and may not justify the expense, although the 1-in-10,000 eruption still could occur. Another factor to consider is the cost-benefit ratio. An evacuation order must also depend upon whether the resources that are saved justify the expenses that are required. Distasteful though it is, a decision not to evacuate because the expense of evacuation cannot be justified could jeopardize the lives of many people.

The number of human fatalities from volcanic eruptions since A.D. 1500 has been large (table 15-2). The data for the period 1600–1986 (table 15-3) suggest that deaths caused by pyroclastic flows, avalanches, and lahars have increased in the twentieth century, but because disasters are measured in terms of lost lives and property, the increase is a measure of population increase. The average number of fatalities per year between 1900 and 1986 is higher than in the preceding three centuries. But after 1900, the numbers of deaths caused indirectly by eruptions, such as posteruption starvation, were significantly less, reflecting the development of rapid relief-delivery systems and global communications.

The previous history of a volcano generally portends its future behavior. Table 15-4 is a list of the criteria that identify dangerous volcanoes. To determine the risks and hazards of a volcano, a score of 1 is given for a statement that is true and a score of 0 for one not true. The higher the total score, the greater the risks and hazards for the volcano.

Despite efforts to predict what volcanoes might do, they often do not follow those predictions, and this failure is a principal difficulty of hazard assessment. There are many factors that affect the prediction of future volcanic activity. For instance, the longer a volcano is inactive, say tens of thousands of years, the greater the amount of rock that is eroded away; therefore, critical geological evidence of past events is destroyed and the ability to forecast behavior based on those events is lost. Another problem is that eruptive behavior may change

Table 15-2. Deaths and Their Causes from Volcanoes since A.D. 1500

Volcano	Country	Year	Pyroclastic Flow	Debris Flow	Starvation	Tsunamis
Kelut	Indonesia	1586		10,000		
Awu	Indonesia	1711		3,200		
Oshima	Japan	1741				1,480
Cotopaxi	Ecuador	1741		1,000		
Papadian	Indonesia	1772	2,960			
Lakagigar	Iceland	1783			9,340	
Asama	Japan	1783	1,150			
Unzen	Japan	1792				15,190
Mayon	Philippines	1814	1,200			
Tambora	Indonesia	1815	12,000		80,000	
Galunggung	Indonesia	1822		4,000		
Nevado del Ruiz	Colombia	1845		1,000		
Awu	Indonesia	1856		3,000		
Cotopaxi	Ecuador	1877		1,000		
Krakatau	Indonesia	1883				36,420
Awu	Indonesia	1892		1,530		
Soufrière	St. Vincent	1902	1,560			
Mount Pelée	Martinique	1902	29,000			
Santa María	Guatemala	1902	6,000			
Taal	Philippines	1911	1,330			
Kelut	Indonesia	1919		5,110		
Merapi	Indonesia	1951	1,300			
Lamington	New Guinea	1951	2,940			
Hibok-Hibok	Philippines	1951	500			
Agung	Indonesia	1963	1,900			
Mount St. Helens	United States	1980	57			
El Chichón	Mexico	1982	>2,000			
Nevado del Ruiz	Colombia	1985		>22,000		

Source: Modified from Tilling (1989), p. 4.

Table 15-3. The Number of Human Fatalities from Volcanic Eruptions between 1600 and 1986

Primary Cause	1600–1899	1900–1986
Pyroclastic flows and debris avalanches	18,200	36,800
Lahars and hyperconcentrated flood flows	8,300	28,400
Ashfalls and ballistic projectiles	8,000	3,000
Tsunamis	43,600	400
Starvation and disease	92,100	3,200
Lava flows	900	100
Lethal gas (carbon dioxide)	—	1,900
Other or unknown	15,100	2,200
TOTAL	186,200	76,000
Average fatalities per year	620	880

Source: Modified from Tilling (1989), p. 5.

with time; therefore, past behavior may not absolutely indicate future activity. And the greater the length of time since the last eruption, the greater the change may be. An important limitation to using past history to assess future hazards is that a volcano may do something entirely new. Another problem is that some hazards are only indirectly related to an eruption and therefore even more difficult to forecast. Such effects include tsunamis, avalanches, and heavy rains that combine with ash to create destructive lahars. Given the present state of knowledge, it is also rarely possible to forecast the magnitude of an eruption, as illustrated by the directed blast of May 18, 1980, at Mount St. Helens. Many scientists thought a lateral blast might occur, but its great force surprised most of them.

Dozens of volcanoes around the world have been assessed with respect to hazards, but many more remain unstudied (table 15-5). The investigations that have been done also vary—from cursory examinations to relatively complete ones—and many of the volcanoes on the list still require more exhaustive research.

Table 15-4. Hazards and Risks

Hazard Rating	Type of Study Needed to Determine the Information
1. Silica content is high. (andesite/dacite/ rhyolite)	Chemical and petrographic research.
2. Major explosive activity within the last 500 years.	Age dating of rocks.
3. Major explosive activity within the last 5,000 years.	Age dating of rocks.
4. Pyroclastic flows or surges in the last 500 years.	Mapping and determining field characteristics.
5. Lahars within the last 500 years.	Mapping and determining field characteristics.
6. Destructive tsunami within the last 500 years.	Field mapping. Age dating.
7. Area of destruction within the last 5000 years is >10 km^2.	Field mapping. Age dating.
8. Area of destruction within the last 5000 years is >100 km^2.	Field mapping. Age dating.
9. Occurrence of frequent volcano-seismic swarms.	Establish network of seismographs.
10. Occurrence of significant ground deformation within the last 50 years.	Establish and monitor ground deformation network.

TOTAL SCORE _____

Source: Modified from Scott (1989), p. 27.

The Lesson of Armero

On November 13, 1985, the Colombian volcano Nevado del Ruiz erupted, depositing ash and producing pyroclastic surges, which melted ice and snow and led to the formation of deadly lahars. One of the lahars flowed through a gorge for about 70 kilometers and emerged from the front of the mountain to overwhelm the town of Armero (see chap. 6).

A little after 9 P.M. local time, seven hours after the eruption began, Nevado del Ruiz ejected pyroclastic flows and surges that scoured and melted part of its glacial cap. The melt water mixed with volcanic debris to create small lahars that became larger as they entered major valleys, where they mixed with the rivers, scraped soil and pore-water from valley sides, and picked up increasingly more debris as they traveled down the valleys. One of the largest lahars to evolve attained a depth of 40 meters and moved as fast as 30–40 kilometers per hour.

Table 15-5. Selected List of Volcanoes that Have Been Assessed for Hazards

Azores	Japan	Philippines
Sao Miguel	Fuji	Mayon
	Asama	Pinatubo
Colombia		
Nevado del Ruiz	Lesser Antilles	Soviet Union
Nevado del Huila	Mount Pelée	Kamchatka
	Soufrière de Guadeloupe	
Ecuador		United States
Cotopaxi	Mexico/Guatemala	Mount Baker
	Tacana	Mount Rainier
Guatemala	Popocatépetl	Mount Adams
Santiaguito	Colima	Mount St. Helens
Pacaya		Mount Hood
	New Zealand	Mount Shasta
Indonesia	Taranaki	Lassen Peak
Merapi	Ruapehu	Mono-Inyo Craters
Kelut	Rotorua	Mount St. Augustine
	Okataina	Island of Hawaii
Italy		
Phlegraean Fields	Papua New Guinea	
Vesuvius	Rabaul Caldera	
Etna	Uluwan	
	Tungurahua	

Note: Relative quality and completeness of studies not given.

The volume of products erupted from Nevado del Ruiz was one-tenth that of the May 18, 1980, eruption of Mount St. Helens, yet 23,000 people died and 10,000 people were left homeless. The economic loss totaled $7.7 billion (U.S.). Of all twentieth-century eruptions, the loss of life was second only to that of Mount Pelée's eruption, in which 29,000 people died.

The 1985 eruption of Nevado del Ruiz will likely go down in history as one of the greatest failures of modern risk mitigation, and most agencies involved bear some of the responsibility. In hindsight, had government officials taken risk evaluations more seriously, the disaster could have been minimized, perhaps averted. Government officials were not willing to bear the economic or political costs of early evacuation or a false alarm. Scientific studies foresaw the hazards, recognizing that many towns along the river courses were vulnerable to lahars generated by snowmelt from eruptions. But scientific reports were not precise enough to guide officials to warn the people in time. According to Barry Voight of Penn State University, the disaster was promoted by the limitations of prediction and detection, the under-advised and inadequately prepared local authorities, the unprepared populace, the refusal to accept a possible false alarm, and the lack of will to act on the uncertain information available. The deadly event occurred only two days before the Armero emergency-management plan was to be critically examined and improved. The numerous misjudgments that delayed progress of emergency management over the previous year, especially the lack of a timely hazard map, contributed significantly to the tragic outcome. It can be said that the catastrophe was caused by an accumulation of human error—not by technological ineptness, not by an overwhelmingly large eruption, and not by simple bad luck.

Intermittent hydrovolcanic activity had begun at Nevado del Ruiz in November 1984 and continued into 1985. On September 11, 1985, the eruptive activity increased and a meeting of national emergency and civil defense groups met in Bogotá on September 17. Colombian volcanologists and foreign scientists produced a preliminary hazard map, which emphasized the hazard from lahars and identified the areas of highest risk, including Armero. Later in the month, Italian volcanologists stated that the monitoring program at Nevado del Ruiz was inadequate and not able to provide rapid warning of changing conditions.

At 3 P.M., on November 13, the eruption began. The civil defense agency in Tolima Province was advised by the Colombian Institute of Geology and Mining (INGEOMINAS) to prepare Armero and other villages along rivers draining Ruiz for immediate evacuation. At 5 P.M., a regularly scheduled meeting of the Emergency Committee of Tolima began about 70 kilometers from Armero. The committee was briefed by INGEOMINAS on the eruptive activity and discussed the evacuation of Armero and other towns and the measures needed to detect possible lahar activity. The police stations in Armero and neighboring towns were alerted. The committee meeting ended about 7:30 P.M., but no firm decisions regarding emergency measures had been made. The Red Cross reportedly recommended limited evacuation of Armero. However, the safety measures, if any, that were taken during the next several hours are not clear from available information. Apparently there was little or no response to the warnings or to the authorities' calls to evacuate.

The main lessons that the Armero disaster taught are that hazard assessments and maps should be prepared well in advance of a volcanic crisis, that the most vulnerable communities should be identified and mitigation measures be taken (public education and a tested warning system) well in advance of a crisis, and that emergency management must make plans for risk communication, uncertainty of events, and false alarms so that decisions are not delayed and rapid public response may occur when there is a crisis. Mitigation at Armero was a failure, but it taught valuable lessons that contributed to the success at Pinatubo.

Mount Pinatubo: A
Success Story

Mount Pinatubo had been quiet for so long that no one could remember it in action, and people lived by the score on its slopes, on its flanks, and along the rivers radiating from its summit. All of that changed beginning on April 2, 1991, when four hundred years of dormancy ended (see chap. 3). Visible eruptive activity began at about 3 P.M. on April 2 with a series of small steam explosions. The eruptions piled thick ash deposits near the vents and severely abraded nearby trees. In the early stages of the eruption, the ejecta consisted of explosively shattered older volcanic rocks; there were no glassy shards, pumice, or bombs from new magma.

The Philippine Institute of Volcanology and Seismology (PHIVOLCS) acted swiftly. On April 5, PHIVOLCS installed a portable seismograph 8 kilometers west of Mount Pinatubo. Within the first twenty-four hours, the seismograph recorded about two hundred small, high-frequency volcanic earthquakes. This sudden and rapid awakening sounded the alarm, and PHIVOLCS immediately recommended evacuation of an area within a 10-kilometer radius around the summit. They installed more portable seismographs 10 to 15 kilometers northwest of Mount Pinatubo and continued to record many very small earthquakes—of Richter magnitudes between 0.5 and 2.5—each day through the end of May. Although the earthquakes were small, a continuous series indicates great accumulative energy release. PHIVOLCS therefore requested and quickly received assistance from the USGS, which dispatched a team of volcanologists to the Philippines on April 22.

From April 29 to May 16, a PHIVOLCS-USGS team installed a radio-telemetered network of seven seismic stations. Data from this network were received and processed at Clark Air Base so that scientists could quickly track changes in earthquake centers and monitor the seismic energy. At this time, the team also prepared a volcanic hazards map, which identified danger zones around the volcano. The map showed that 600 years ago, pyroclastic flows had reached the area of present-day Clark Air Base and the now densely populated areas south and east of the volcano. Voluminous pyroclastic flows had also swept through the region 2,500 and 4,600 years ago (see fig. 3-9).

Revelations about Mount Pinatubo's violent and distant past and the onset of the new activity put the volcanologists in the spotlight—a difficult position of responsibility. A wrong decision could cost thousands of lives or result in the huge expenses of an unnecessary evacuation. But the evacuations at Mount Pinatubo would be successful because of effective communication between the volcanologists and the public officials, who then relayed the information to the public. The volcanologists at the volcano developed a simple scheme consisting of five levels of emergency responses (table 15-6). The schedule was

Table 15-6. Emergency Response Levels Used for Evacuation Warnings at Mount Pinatubo

Alert Level	Criteria	Interpretation	Date Declared
No alert	Background quiet	No eruption in foreseeable future	n/a
1	Low-level seismic, fumarolic, unrest	Magmatic, tectonic, or no eruption imminent	n/a
2	Moderate level of seismic, other unrest, with positive evidence for involvement of magma	Probable magmatic intrusion; could eventually lead to an eruption	5/13/91
3	Relatively high and increasing unrest including numerous b-type eqs., accelerating ground deformation, increased vigor of fumaroles, gas emission	If trend of increasing unrest continues, eruption possible within 2 weeks	6/5/91
4	Intense unrest, including harmonic tremor and/or many long-period (= low-frequency) earthquakes	Eruption possible within 24 hours	6/7/91 5:00 P.M.
5	Eruption in progress	Eruption in progress	6/9/91 5:15 P.M.

Note: Reverse procedures as eruption wanes: To protect against "lull before the storm," alert levels are maintained for the following periods *after* activity decreases to the next lower level. From level 4 to 3, wait 1 week; from level 3 to level 2, wait 72 hours.

distributed to civil defense and local officials beginning on May 13, when the volcano was at level 2.

The centerpiece for briefing local and national officials and citizens on possible activity was a videotape about pyroclastic flows and lahars, hazards that had been generated by the volcano in the past. It explained the volcanic hazards without geological jargon, making it readily understood. Ironically, it had been prepared by the French volcanologists Maurice and Katia Krafft, who would lose their lives on June 3, 1991, at Unzen Volcano, Japan, from a pyroclastic surge at the very same time that their video was being used to save thousands of lives in the Philippines.

The newly established seismic network showed that volcanic earthquakes were originating about 5 kilometers northwest of, and from 2 to 6 kilometers below Mount Pinatubo's summit. There were few earthquakes beneath the steam vents at the summit, but an instrument called a correlation spectrometer (COSPEC), which measures sulfur dioxide, showed a ten-fold increase in its emission between May 13 and May 28. The volcanologists interpreted this to mean that juvenile (new) magma was rising toward the surface. Through most of May, gases from the summit indicated an imminent eruption, but the earthquake centers did not move upward; however, in late May and the first few days of June, they moved to the region below the summit steam vents. Harmonic tremors produced by liquids moving through subterranean passageways showed that magma was on the move.

On June 6 and 7, there was an intense swarm of shallow earthquakes directly beneath the vent area, and it was accompanied by swelling of the upper east flank of the volcano—another warning that magma was moving. A thick lava dome then expanded into the crater and was followed by continuous explosive eruptions of ash. After the extrusion of the dome, the earthquakes became concentrated beneath it. By June 11, it had nearly doubled in size, and shallow earthquakes of increasing intensity were occurring.

The signs were clear. Mount Pinatubo was almost certainly building up to larger eruptions! The alert level was raised to 3 on June 5, to 4 on June 7, and to 5 on June 9. The radius of evacuation was increased to 15 and then to 20 kilometers, and an additional 20,000 people were moved to evacuation camps. Aircraft were moved from Clark Air Base, and on June 10, 14,000 U.S. military personnel and their dependents were evacuated.

Shortly thereafter, on June 12 and 13, a series of powerful explosions expelled ash columns—first to 19,000 meters above sea level, then to over 24,000 meters. The eruptions of June 12 prompted the radius of evacuation to be increased to 30 kilometers, which raised the number of evacuees to 58,000. The explosions destroyed part of the new dome and left a crater 200–300 meters in diameter southeast of the dome. Pyroclastic flows swept down all sides of the volcano to a maximum distance of 6 kilometers during this explosive phase. Coarse ash fell on communities to the southwest, west, and northwest.

The evacuations were accomplished just in time, for the main eruption took place on June 15. A lateral blast at about 6 A.M. exploded outward in all directions, accompanied by an eruption column that rose to 12,000 meters above sea level and spread out as a broad collar around the volcano. Seven or more subsequent eruptive pulses evolved into essentially one continuous strong eruption by early afternoon. At about 1:45 P.M., the frequent seismic signals so interfered with one another that they were virtually unreadable. Within twenty-five minutes all seismic stations except at Clark Air Base were knocked out. The base was in total darkness as ash filled the sky. The eruption was most intense between 3 and 4 P.M., when satellite images documented the ash plume to be 35–40 kilometers high.

Pyroclastic flows swept over an area of roughly 100 square kilometers to blanket nearly all the areas of prehistoric pyroclastic flows shown on the hazards planning map. Pyroclastic flow deposits accumulated up to 200 meters in deep canyons close to the volcano. Many canyons were filled so completely that the new ground surface around the volcano formed a nearly flat, featureless plain. One of the longest pyroclastic flows passed within 200 meters of Clark Air Base, extending 16 kilometers from the old summit and 4 kilometers short of the longest known pyroclastic flow on Mount Pinatubo's east side.

An unanticipated hazard resulted from typhoon rains, which saturated the loosely deposited ash, doubled its weight, and caused many roofs to collapse. The rain-saturated ash on the volcano turned into lahars that caused great damage along all the major stream valleys. The lahars were the greatest hazard of the eruption, and by June 16, about two hundred thousand people, in addition to those who had been already evacuated, fled their valley homes for refuge in evacuation camps—the largest evacuation ever prompted by a volcanic hazard.

The volume estimates of fallout ash and pyroclastic flow deposits from Mount Pinatubo are between 3 and 7 cubic kilometers. (Pyroclastic material has about half the volume of magma.) If these volume estimates of ash are correct, the eruption of Mount Pinatubo was one of the largest eruptions of this century, second only to that of Novarupta (Katmai), Alaska, in 1912, which produced the Valley of Ten Thousand Smokes. The effects of such large eruptions are felt for decades thereafter.

In 1991, the rainy season began in July but worsened in August, September, and October, causing dangerous lahars. By December 1991, nearly every bridge within 30 kilometers of Pinatubo had been destroyed, several lowland towns were partly buried, and most other lowland towns within 50 kilometers of the volcano were threatened by floods. As little as 1 centimeter of rain within thirty minutes can saturate ash and trigger new lahars—rain occurred daily in August and weekly in December 1991. Because pyroclastic flow deposits fill valleys, rivers are choked with debris, increasing their volume and erosive capability. Downstream channels are filled with sediment. Lahars spill over the banks of filled channels and cover the surrounding countryside. Blockage of smaller tributaries causes new lakes to form behind natural dams; when the dams break, there is sudden flooding. The studies made before the great eruption of June 15 created a foundation of knowledge that saved thousands of lives. Fifty-eight thousand people were eventually evacuated during the first weeks of the eruption. Most were saved, but 320 people perished from sickness in refugee camps and from the collapse of ash-laden roofs.

As of February 1996, the eruption was essentially over, but lahars still continue to plague the region. The mountain has become a store of loose sediment that will feed future lahars for many years. At least 30 percent of the total ash volume is expected to be moved by natural processes in the next five to ten years.

Table 15-7. Recent Deaths of Volcanologists

Name	Volcano and Year
Robin Cooke, Elias Ravin	Karkar Volcano, Papua, New Guinea, March 8, 1979
David Johnston	Mount St. Helens, Washington, USA, May 18, 1980
Harry Glicken, Katia Krafft, Maurice Krafft	Mount Unzen, Kyushu, Japan, June 3, 1991
Néstor García, Fernando Cuénca, José Arlés Zapata, Geoff Brown, Igor Menyailov, Carlos Trujillo	Galeras Volcano, Colombia, January 14, 1993
Victor Hugo Pérez, Alvaro Sánchez	Guagua Pichincha Volcano, Ecuador, March 12, 1993

Source: Data from the Smithsonian Institution.

Volcanologists on the Front Lines

While studying volcanoes and working to make life safer for people living near active volcanoes, volcanologists sometimes put themselves into dangerous situations. There is always the possibility of a surprise eruption, although the odds are small that one will occur without warning. Many volcanologists have a jet-passenger philosophy: they know that eruptions happen infrequently and gamble that one won't happen when they're on the volcano. Volcanologists are sometimes portrayed as daredevils who push their luck as they study exploding volcanoes (see chap. 7), but most are ordinarily cautious individuals with an extraordinary curiosity and passion to learn about the workings of the earth and of volcanoes in particular. Few volcanologists take foolish chances that add to an already dangerous situation. Still, a few volcanologists have died while conducting research on active volcanoes. During the first half of the 1990s, coincidentally the Decade of Natural Disaster Reduction, eleven volcanologists died while collecting data near or on volcanoes (table 15-7).

The Tragedy at Galeras Volcano

On January 14, 1993, Galeras Volcano, located near Pasto, Narino Province, Colombia, became a killer. Galeras is a Decade Volcano, and from January 11 to 16 a workshop of volcanologists was held in Pasto to study Galeras first-hand. Field excursions were planned for Thursday, January 13, but were changed to the following day because of electrical shut-downs. By pure chance, Friday became the unlucky and tragic day, for at about 1:40 P.M. local time, an unanticipated volcanic explosion killed six of the volcanologists and three tourists.

Mario Mazzoni, a volcanologist from the Centro de Investigaciones Geológicas, La Plata, Argentina, who was with the field workshop on the flank of Galeras at the time of the explosion, relates the impressions of the volcanologists who were a few kilometers from the crater when the explosion occurred.

I had been taking pictures and stopping in different places. The outcrops were plenty and the group stretched along several hundred meters. At the moment of the explosion I was alone taking pictures in the most advanced position, or in other words, lower in the valley in relation to colleagues in the group.

I felt what at first sounded like a thunderstorm. It did not impress me because it was cloudy, as all the morning had been. But the noise continued much longer than thunder usually does, maybe more than fifteen seconds. My perception was also that the rumbling noise was the result of thousands of smaller noises which swiftly came toward us down the valley by the canyon.

I, and the rest of the group then suspected that Galeras was erupting. I realized then that we were in a dangerous place, as the Azufral Canyon is the natural channel for any output of the Volcano and also that the rumbling noise could be ascribed to a mass flow descending from the top of the mountain. My instinctive reaction was to walk quickly up, coming back by the road to get to a higher and safer topographic position in the narrow valley, and to join the other members of our field workshop group, to check their impressions.

The same sensations were experienced by the majority of colleagues. And though we wondered how the echoed thunderstorm would sound in a canyon like Azufral, each intimately believed an eruption of Galeras had occurred. We all knew that at the time we felt the noise, other groups may have been in the crater or pretty near to it. We were very upset and shocked by this possibility. A farmer in the canyon told us that the noise was not a thunderstorm and that it had sounded like in previous eruptions of Galeras.

. . . The group decided that the excursion was over, and to return to Pasto by the shortest road. While coming back by the southeastern flank of the volcano it began raining. Wipers collected fine ash with water in the lower part of the windshield. Unfortunately it was the proof of what we feared more, and anticipated the concern and distress we felt on arriving in Pasto, where it was confirmed that it had been a tragic eruption. (quoted in IAVCEI, *Commission on Volcanogenic Sediments*, p. 4)

The summit of Galeras, 4,270 meters above sea level, is in southwestern Colombia near the Ecuadorian border (fig. 15-4). It reawakened in 1988 after forty years of inactivity and in 1991 was chosen as one of the Decade Volcanoes because it is active and lies only 6 kilometers from Pasto, a city of about three hundred thousand people. The precarious position of Pasto makes it mandatory that the residents always be prepared for a possible evacuation.

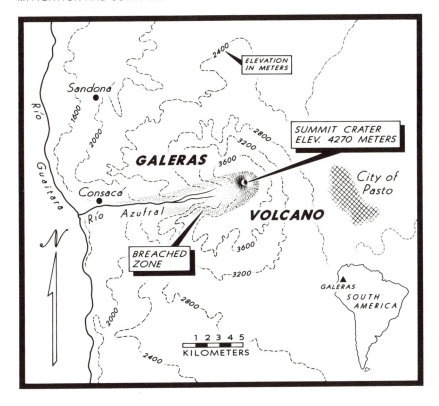

Fig. 15-4. Location of Galeras.

The volcano has erupted at least six times in the past 4,500 years, producing small pyroclastic flows, and Pasto is well within the area that these flows covered. There was minor gas activity inside the crater in April 1988, and a gas plume rose in February 1989. This first explosive activity visible from Pasto alerted the local authorities to begin publicizing the possible consequences of future eruptions, and a USGS workshop brought volcanologists together to study deposits produced by recent eruptions, to install new instruments, and to enhance the capabilities of the new Observatorio Volcanológico del Sur. Explosive activity prompted INGEOMINAS to install the first of several seismic stations on the flanks of Galeras. INGEOMINAS had made remarkable progress in responding to the challenges of volcanology in Colombia after the 1985 disaster at Nevado del Ruiz.

The juxtaposition of a large city and a newly reactivated volcano with a past history of pyroclastic flows prompted INGEOMINAS and the National

Disaster Prevention Office to request Decade Volcano status as part of the United Nations' IDNDR. Thus, INGEOMINAS, in cooperation with Professors Stanley Williams, from Arizona State University, and John Stix, from the University of Montreal, organized an international workshop in Pasto, Colombia beginning January 11, 1993. There were 101 participants from fourteen countries. The purpose was to inspire participants to propose geological, geophysical, and geochemical research programs to improve knowledge about the volcano and increase understanding of its hazards. Another important purpose was to foster better understanding between volcanologists, local officials, and residents.

Six excursions from the workshop were held on January 14. Several groups visited the summit area on Thursday morning and then returned to Pasto as planned, while others remained in the area near the crater. The organizers decided to take a few volcanologists into the crater for gas sampling. The risk of eruption was considered to be low because following an explosion on July 16, 1992, the seismicity of Galeras had gradually declined to fewer than ten events per day. From early- to mid-January 1993, there had been no significant changes in activity, except for several minor earthquakes beginning two weeks before the workshop. Also, no changes were detected on two tiltmeters installed on the flanks of the volcano, and the last fieldwork done in the crater on December 16, 1992, reported no changes in volcano status. Two other indications of inactivity were that the temperature of a fumarole had decreased and that the emission of sulfur dioxide gas was low.

The group of twelve volcanologists led by Professor Stanley Williams, accompanied by several tourists and journalists, descended into the inner crater to sample gases and take microgravity measurements. Most of the work was done by noon, and as the volcanologists were preparing to leave the crater, the ground suddenly began to move upward, followed by an explosion that produced a rain of rocks that were up to a meter in diameter. Six scientists—Néstor García, Fernando Cuénca, José Arlés Zapata, Geoff Brown, Igor Menyailov, and Carlos Trujillo—and three accompanying tourists were killed in the explosion. Four other scientists were injured—Mike Conway, Luis Lemarie, Andy Mcfarlane and Professor Williams himself who later described what happened:

> By about 10:30 a.m., six of us were working on the rim of the crater at the Deformes fumarole, which released gas about 220° C. By 1 p.m., I was encouraging people to climb down to the moat separating the volcano and the caldera's inside scarp and to climb out to the caldera rim. Igor Menyailov (Institute of Volcanology, Petropavlosk, Kamchatka) and Néstor García (National University) were in the crater collecting gas samples released at 600° C from the margin of the remaining part of the dome. Geoff Brown (Open University), Fernando Cuénca (Bogotá) and Carlos Trujillo (Pasto) were on the western rim of the crater to assist their research and to

make sure that they safely exited. Experience at many volcanoes had taught me that clouds and fog can arrive rapidly and create serious problems for anyone without detailed knowledge of a volcano's physiography.

1:43 p.m.—As I talked with Igor, who had just finished collecting samples, rock fall occurred on the walls of the crater. I yelled to everyone that this was a bad sign and that we should escape as fast as possible. Before anyone had a chance to react, the first explosion blew such large blocks at such high speeds and low lateral angles that all those in the crater or on the rim (except me) were killed. I turned and began to run down (as were those who had been somewhat below the rim). I made it only 10 meters before more debris killed four more people (José Arlés Zapata, a Pasto Observatory scientist, and three tourists who just happened to be there), and gave me my first injuries (a skull fracture and a broken jaw). I got up and ran about the same distance before the next impact shattered the bones in the calf of my right leg (and broke my left calf, less critically). I got up and tried to run, only to fall because of the compound break. The others, who were close (Luis Le Marie [*sic*], of Escuela Poli-técnica Nacional in Ecuador; Andrew Macfarlane [*sic*], of Florida International University; and Michael Conway, of Michigan Technological University) and just a short distance down the slope of the volcano, were also injured, but all were able to continue moving down toward the caldera scarp (a much safer distance from the crater). Andrew Adams, of Los Alamos National Laboratory, and Alfredo Renée Roldán, of Guatemala, were already on the caldera scarp and so suffered much less. I crawled to a large boulder, which I knew would provide some shelter, and watched for additional bomb impact so that I could roll out of the way. I was on fire because the incandescent blocks had ignited my backpack and clothes (rolling around stopped the flames).

After about 15 minutes of ash fall, the eruption stopped, and terror, tragedy, and the uncertainty of what lay ahead took control. I lay there for about two hours before hearing two incredibly brave women, Marta Calvache and Patty Mothes (who lives in Ecuador, where she studies volcanoes with her husband, Pete Hall), calling out to survivors in an effort to organize rescue efforts. I was the last survivor to be carried out by them (with the help of others) because the distance to me and the nature of my injuries made it more difficult to carry me out. (Williams, "Perspectives," p. 13)

Andrew Adams of the Los Alamos National Laboratory was on the caldera scarp, and was hit in the head by a large block. He was the only member of the group that wore a protective helmet and body pads, and they saved his life. No one else had worn protective gear, and another lesson was learned the hard way. The terrible experience led to a statement of precautions written by the International Association of Volcanology and Chemistry of the Earth's Interior. The essence of those rules lies in being well prepared, and above all, wearing protective gear.

The eruption of Galeras underscores the capriciousness of volcanoes, for it exploded with little warning. Fortunately, most volcanoes give sufficient warning of impending danger with increasing seismic activity, ground deformation, rising hot-spring or fumarole temperatures, and increasing emission of sulfur dioxide and other volcanic gases.

Dante and the Volcano

Appearing to be straight from a science fiction film about a Martian invasion of the earth, a spidery figure—Dante I, an experimental robot—moved in slow motion across land so rugged that no lunar rover could travel. Dante I failed to successfully negotiate rough land, but the lessons learned from its failure were applied to Dante II, which met with modest success. Dante II entered the crater of the Alaskan volcano Mount Spurr in the summer of 1994 and spent seven days in the active crater to gather gas-emission data in the inhospitable environment of hot fumaroles and quaking ground. The robot was tethered to a fixed portable generator and carried six video cameras and gas and temperature sensors. It was programmed to function on its own, using information from a scanning-laser range finder to build a virtual terrain map to chart its own course. Dante II successfully recorded the data and transmitted it to outside computers. On its way out of the crater, however, the robot tipped over when one of its legs sank into a soft spot. Initial rescue attempts caused it to tumble another 3 meters toward the crater floor. Two people had to risk their lives to go into the dangerous environment and hook Dante II onto the cable dangling from a helicopter, which then lifted the robot out to the rim.

Could Dante II have been used to collect gas samples in the crater of Galeras Volcano and so have saved the six volcanologists who died? Given even its present, early stage of development, it most likely could have been sent into the crater to determine the type of gas emissions that the volcanologists were measuring at the time of the tragedy. The namesake of the fourteenth century poet, Dante is being prepared for the remote exploration of rugged inhospitable solid planets and moons within our solar system. What better place to test Dante for exploration of planets than in the dangerous environment of an active volcano? The spidery robot cannot withstand lava lakes or survive a direct hit by flying projectiles of lava and falling rocks, but it can collect information in dangerous places. Someday, Dante may be able to substitute for living volcanologists in places where they cannot go and where critical information needs to be gathered. But the present $1.7 million price tag prohibits the robot's widespread use; it is also slow-moving and complicated to use, and the terrain of active volcanoes is treacherous and changing. In an active volcano, Dante needs to be rugged, quick, nimble, and "smart." In the hot or freezing, airless static environment of other planets, Dante II and its progeny will certainly be powerful data-gathering tools, but for volcano research, Dante is currently a very crude pioneer.

Even though Dante is presently unsuitable for rugged duty in a volcano, its experience in and on volcanoes has led to improvements that will enhance its use in extraterrestrial exploration. Dante II may eventually spawn robots rugged, simple, and inexpensive enough to be useful in gathering data within active volcanoes, and along with collecting remotely sensed data on temperatures and gases, they may possibly save lives of future volcanologists.

References

Aramaki, S., F. Barberi, T. Casadevall, and S. McNutt. "Report of the IAVCEI Sub-Committee for Reviewing the Safety of Volcanologists." *WOVO [World Organization of Volcano Observatories] News* no. 4 (autumn 1993): 12–14.

Crandell, D. R., and D. R. Mullineaux. *Potential Hazards from Future Eruptions of Mount St. Helens Volcano, Washington.* USGS Bulletin 1383-C. Washington, D.C., 1978, pp. 1–26.

Gore, Rick. "The dead do tell tales at Vesuvius." *National Geographic*, May 1984, 557–613.

Gunther, Judith Anne. "Dante's Inferno." *Popular Science*, November 1994, 66–68, 96.

International Association of Volcanology and Chemistry of the Earth's Interior (IAVCEI). "CVS Members Experience Fatal Eruption of Galeras Volcano, Colombia, January, 1993." Commission on Volcanogenic Sediments Newsletter, no. 8 (August 1993): 2–4.

Luhr, J. F. and J. C. Varekamp, eds. "El Chichón Volcano, Chiapas, Mexico." *Journal of Volcanology and Geothermal Research* 23 (1984): 1–191.

Nakamura, K., "Volcano-stratigraphic study of Oshima Volcano, Izu." *Bulletin of the Earthquake Research Institute* 42 (1964): 649–728.

Muños, F. A., et al. "Galeras Volcano: International workshop and eruption." *EOS* 74 (1993): 281–87.

Pinatubo Volcano Observatory Team. "Lessons from a major eruption: Mt. Pinatubo, Philippines." *EOS* 72, no. 49 (December 3, 1991): 545, 552–53, 555.

Pringle, Patrick T. *Roadside Geology of Mount St. Helens National Volcanic Monument and Vicinity.* Washington Department of Natural Resources Circular 88, 1993.

Schofield, Julie Anne. "Dante Survives the Inferno." *Design News* (September 26, 1994): 69–74.

Scott, W. E. "Volcanic-hazards zonation and long-term forecasts." In *Volcanic Hazards*, edited by R. I. Tilling, 25–49. Washington, D.C.: American Geophysical Union, 1989.

Tilling, R. I., ed. *Volcanic Hazards.* Short Course in Geology: Vol. 1. Washington, D.C.: American Geophysical Union, 1989.

Voight, B. "The 1985 Nevado del Ruiz volcano catastrophe: anatomy and retrospection." *Journal of Volcanology and Geothermal Research* 44 (1990): 349–86.

Williams, S. N. "Perspectives from a researcher and survivor." *Geotimes* 38 (1993): 12–14.

Wright, T. L. and T. C. Pierson, *Living with volcanoes.* U.S. Geological Survey Circular 1073. Washington, D.C. 1992.

The Volcano Traveler

For the curious traveler who has never visited a volcano, active or inactive, we give brief summaries of a select few to visit. The list is necessarily incomplete partly because we include only those that are reasonably accessible to the average tourist; nevertheless, it would take several years to see them all. Rather than providing sketch maps here, which can become rapidly obsolete, we advise you to purchase local maps before your visit or after you arrive. It is hoped that our guide will inspire many of you to visit a volcano even if it is not included here. Some volcanic areas not included here are: the Azore Islands, Canary Islands, central France, central Germany, and the Aleutian Islands.

A Word of Caution

Inactive volcanoes offer beautiful outdoor recreational experiences similar to nonvolcanic recreation sites. Visitors must use common sense when utilizing their beauty and outdoor challenges. Don't overextend yourself, carry plenty of water, and obtain a trail guide.

Active volcanoes are a different story. Most outdoor sites wait for you to visit, but an active volcano sometimes will come to you. Use extreme caution! Before going anywhere near the activity, or even if the volcano is not showing any activity during your visit, consult the local experts on site at a volcano observatory. If there is no observatory, consult the local police or park rangers about the nature of the activity and safe places from which to observe it. Follow their recommendations, requests, or regulations for they know the dangers of their volcano.

Visits to volcanoes are to be enjoyed—the trips are always worth it, but the awesome power of volcanoes requires you to respect that power. Use common sense and the advice from local experts during your visit, and be aware of the dangers noted below.

(1) **Lava flows**. Not all lava flows are red-hot running lavas; some are very slow-moving slags that seem solid and black on the surface but are liquid beneath it. Stay off active flows no matter how benign they seem; watch them from the side, from a hill, or from an aircraft. Do not attempt to cross an active flow. Do not allow yourself to be trapped by multiple flows.

(2) **Pyroclastic flows**. If an explosive eruption is producing pyroclastic flows, do not

visit the volcano. Pyroclastic flows can travel 100 kilometers or more. There is no escape from these flows; if the blast doesn't kill you, the high temperatures will.

(3) **Volcanic domes**. Slowly growing volcanic domes and plugs in craters may seem benign, but most of the volcanologists and nature tourists killed during the last ten years have been in situations in which a dome exploded without warning. Ask the local experts if the dome has been dormant for over ten years. Even then be careful, because footing is treacherous, and the glassy rocks can be as sharp as razors.

(4) **Lahars and floods**. In valleys leading from erupting volcanoes where there has been a history of lahars and floods, use caution when crossing streams or selecting a campsite in low-lying areas. Never cross a flowing lahar even if it is very small, for debris flows often move in small and large pulses.

(5) **Gases**. Closed basins (for example, craters) in areas where there is release of volcanic gases should be avoided. Never enter a basin or depression in areas where there is a history of abundant volcanic gas release. Gases such as carbon dioxide, sulfur dioxide, and hydrogen sulfide are heavier than air and can collect in pockets to kill quickly and silently if the level of the gas pool is higher than your nose. You may not be able to hold your breath long enough to escape, and neither may potential rescuers. If you encounter a crater or depression where there is dead vegetation, carcasses, or many animal bones, do not enter it. Be particularly careful on windless days, when gases are not dispersed.

(6) **Geothermal areas**. Hot springs, bubbling mudpots, and geysers are fascinating. However, do not venture across unexplored ground if there are many of them. In explored areas, stay on designated trails. Within these hydrothermal areas, the thin crusts of silica sinter that hide pools of boiling water may break if stepped upon. Falling into such pools will boil you alive or, at the very least, cause third-degree burns.

Canada and United States

Mount Garibaldi, British Columbia

One of the northernmost of the Cascade volcanoes, Mount Garibaldi, is only 65 kilometers (40 miles) north of Vancouver, British Columbia, and is the centerpiece for Garibaldi Provincial Park. Although long inactive, this glacier-covered volcano is one of the youngest in Canada.

Getting there: Drive north about 80 kilometers (50 miles) from Vancouver on Highway 99 past Squamish. Watch for signs to Diamond Head Base Camp. All access to the park is on foot or cross-country skis (winter).

Note Well: The weather can range from warm to dangerously cold. Prepare for rain, snow, and sleet. Because this is a wilderness area, carry adequate food and water.

Mount Baker, Washington

One of the lesser-known of Washington's youngest stratovolcanoes, Mount Baker has been active intermittently over the last ten thousand years. The most common activity has been lahars that originate on its slope and travel down the valleys. Increased fumarolic activity in 1975 created anxiety in communities along the rivers that originate on its slopes. It was thought that landslides would cascade into a lake, displace water, and cause a dam to overflow and flood towns downstream.

Getting there: For an overview, take Highway 542 east from the city of Bellingham for about 50 kilometers (31 miles) to Glacier Ranger Station. About 0.8 kilometers (0.5 miles) east of the ranger station, turn south on Glacier Creek Road (Forest Road 3904). Drive 13 kilometers (8 miles) to a turn-around viewpoint.

For a view from the south, drive north from the town of Concrete on the Baker Lake Highway (on Highway 20, the North Cascades Highway) to 0.8 kilometers (0.5 miles) beyond Baker Lake Resort. Turn left on Shuksan Lake Road (No. 394) and follow it to its end.

Note well: Mount Baker is covered with glaciers and should be climbed only by expert mountaineers. If you are not satisfied with views from a distance, discuss your trip with the Forest Service or one of the mountaineering organizations in Washington.

Mount Rainier National Park, Washington

Mount Rainier is the Decade Volcano for the mainland United States that was chosen for detailed study during the International Decade of Natural Disaster Reduction (see chap. 15). Mount Rainier, with its summit at 4,392 meters (14,410 feet), looms over the sea-level cities along Puget Sound. It is draped with twenty-six glaciers and contains the headwaters of many of the rivers draining into the sound. Throughout its history, there has been constant interaction between eruptions and ice; the last explosive eruption was in the mid-1800s, and more than sixty debris avalanches and lahars have swept the valleys around the mountain during the last ten thousand years. Today the volcano is not entirely asleep. It has summit fumaroles that have altered the rocks over 1 square kilometer, where climbers who have reached the summit can capture some warmth. It is a volcano to be admired, yet carefully watched.

Getting there: The most popular route is a loop, beginning in the Tacoma area, along Highway 7 south from Parkland, for 54 kilometers (34 miles) to Elbe. The road runs east as Highway 706 for 32 kilometers (20 miles) to Longmire, which is in the park. The road continues along the southern slope of the volcano until it intersects Highway 123; this highway goes north, across Cayuse Pass (closed in winter), to its intersection with Highway 410, which runs along the northeastern park boundary and loops back toward Tacoma (about 112 kilometers; 70 miles). Roads within the park go to a variety of viewpoints and campgrounds.

Note well: Weather in the high mountains is notoriously unpredictable and dangerous. Hikers should stay on marked trails and carry foul-weather clothing and food. Off-trail climbing should only be done with experienced guides.

Mount St. Helens Volcanic Monument, Washington

Mount St. Helens, once a graceful and symmetrical stratovolcano with an elevation of 2,951 meters (9,677 feet), was reduced by collapse of its summit to 2,551 meters (8,365 feet) and erupted 2.5 cubic kilometers of debris (see chap. 1). The collapse created the first witnessed debris avalanche. The removal of large amounts of rock from the volcano by the avalanche reduced the pressure on the cryptodome within the volcano's internal summit region and caused an explosion that projected a blast northward, knocking down forests and killing fifty-seven people. Within five to ten minutes, it created 572 square kilometers (203 square miles) of desert-like landscape that is now being encroached by fireweed and other small plants. A north-south swath of land that encompasses untouched forest on the north and south and devastated land in its central part was dedicated as a National Volcanic Monument in 1982 by an act of Congress and is managed by the U.S. Forest Service. The monument was created to protect the unique environment formed by the eruption, to provide a site for scientific research, and to be an educational resource about volcanoes for the public. There are lakes for fishing, many kilometers of hiking trails that are well labeled with information markers, and unsurpassed views of a recently active volcano.

Getting there: Buy a map of Mount St. Helens National Monument or obtain the guidebook to Mount St. Helens written by Patrick T. Pringle. This small but very informative book gives a history of Mount St. Helens and the eruption of 1980 and a running commentary of the geology of the routes that you may travel.

For viewing the volcano, there are two major routes, one from the west and one from the east.

Route from the west: Take U.S. Interstate Highway 5 to Castle Rock and turn eastward on State Highway 504, which goes into the heart of the devastated land north of Mount St. Helen's crater. At 8.5 kilometers (5.3 miles) from Interstate 5 is the Mount St. Helens Visitor Center with excellent exhibits and a well-stocked bookstore on nature topics. Continue eastward past Toutle (last chance for gasoline). The recently constructed highway has many scenic turnouts. It ends at Coldwater Ridge Visitor Center (70 kilometers; 43.5 miles from Interstate 5), with its unobstructed view of Mount St. Helens and the dome within its crater. Currently under construction is a road from the visitor center to the top of Johnston Ridge (80.5 kilometers; 50.3 miles), directly north of the crater across the Toutle River. At the end of the road is the proposed Johnston Ridge Observatory, located near the observation post where USGS geologist David Johnston lost his life on the morning of May 18, 1980.

Route from the east: There are two main ways to the eastern side of the devastated area, both leading to paved Forest Road 99 which leads through the eastern side of the devastated area. One route begins by turning eastward from Interstate 5 onto State Highway 503 at Woodland. Travel through the small town of Cougar and continue toward Forest Road 99. Be sure to fill your gas tank at Cougar. Follow the road signs toward Randle. The distance from Interstate 5 to Forest Road 99 is 117 kilometers (72.7 miles). The second approach is Highway 12 from eastern or western Washington, which runs through Randle. From Interstate 5 on Highway 12, the distance to Randle is 78 kilometers (49 miles). Turn south on Highway 131, which becomes Forest Road 25 (paved). Gas up at Randle if you continue south because there are no stations until you reach Cougar, about 160 kilometers (100 miles) away. Forest Road 99 is 34 kilometers (21 miles) south of Randle and is a two-lane paved road with many well-marked turnouts and explanatory signs. There are many well-marked trails. The road ends at Windy Ridge, which is a trailhead into the monument and for many kilometers of trails. From Windy Ridge is an unobstructed view of the Toutle River Valley to the west, but you cannot see into the crater of Mount St. Helens from Forest Road 99.

Note well: Be alert while traveling the narrow, winding roads, and fill up your gas tank at local towns, for gasoline stations are few in this region. The monument is a natural laboratory and exhibit. Respect the landscape and scientific plots, equipment, or experiments that you may find. Stay on marked trails and do not remove ash, pumice, rock, or plant samples from inside the national monument without a permit. Pack your litter out. On hikes be sure to take plenty of water.

Mount Hood, Oregon

Take time to visit Mount Hood and its many volcanological features. The route from Portland to central Oregon crosses the southern flank of this 3,427 meter high (11,245 feet) stratovolcano with its beautiful views. A venerable mountain inn, Timberline Lodge, sits at an elevation of 1,828 meters (6,000 feet) and is a trailhead for climbers and the focus for summer skiing. Mount Hood has erupted many times in the past, the last being in 1865. Detailed hazard assessment is currently being done for the volcano.

Getting there: Take Highway 26 (the highway to Bend) east from Portland, through the towns of Greshan and Sandy. About 89 kilometers (55 miles) east is the settlement of Government Camp, where a side road goes north a couple of miles to Timberline Lodge. Highway 35 branches north, along the eastern slopes of Mount Hood, to intersect Interstate Highway 80 near Hood River. Stop at the visitor center between Rhododendron and Government Camp for the museum, maps, and brochures, for there are many forest roads on the slopes of Mount Hood.

Note well: Do not consider climbing to the summit without expert advice or guidance. The weather can turn bad in less than an hour.

Three Sisters / Mackenzie Pass, Oregon

The ragged surfaces of young aa lava flows can be experienced at Mackenzie Pass across the Cascade Mountains. The highway cuts through young, rubbly lava flows from Belknap Crater and Yapoah Cone—a vast, unexpected clearing in the Cascade Mountain forest. An observatory along the highway built from the stone of the lava flows on which it rises provides impressive views of Mount Jefferson, a stratovolcano, numerous cinder cones, and the North and Middle Sisters. The Three Sisters are on a line with Broken Top volcano. They are young, glaciated stratovolcanoes along the spine of the Cascade Range.

Getting there: Take Highway 126 east from Eugene/Springfield for 100 kilometers (60 miles) to Belknap Springs, where Highway 242 leaves the highway toward Mckenzie Pass and the town of Sisters. Highway 242 climbs for 35 kilometers (22 miles) through the beautiful, heavily wooded mountains of the western Cascade Range. It breaks into open ground at the pass, where the shrubbery and trees have been eliminated by lava flows and are not yet reestablished. Highway 242 continues east for another 24 kilometers (15 miles) to the tourist town of Sisters on the eastern Cascades slope, where it rejoins Highway 126.

Note well: Highway 242 is closed during the winter; ask about local road conditions before starting out in the winter months.

Newberry Volcano National Monument, Oregon

Close to a main highway (Highway 97), this volcano has a great variety of volcanic features: a broad lava shield with hundreds of cinder cones on its outer slopes, a magnificent caldera (without the hordes of tourists that plague Crater Lake), an obsidian flow, a pumice cone, a tuff ring, and twin lakes, complete with trout. A magnificent view is the reward for taking a graveled road that turns south from the monument entrance and climbs to the top of Paulina Peak, part of the precipitous southern caldera rim. A large part of central Oregon's volcanic scenery can be seen on a clear day from Paulina Peak, but best is the overview within the caldera with its maar volcanoes, explosion craters, obsidian flow, and pumice layers. The various volcanic features are accessible by forest roads, paved roads, and well-marked interpretive trails.

Getting there: Travel south from Bend along Highway 97 for 48 kilometers (30 miles). Along the highway is Lava Butte, a cinder cone where a shuttle bus will drive you to the top to view its crater, and Lava River Cave, where you can explore a lava tube. The turnoff to Newberry Volcano with its Paulina Lakes is several kilometers north of the town of LaPine.

Note well: Plan on several days to explore this area. The weather is delightful but cold and snowy during winter, since it is beyond the rain shadow of the Cascades.

Crater Lake National Park, Oregon

Crater Lake, the youngest caldera in the lower forty-eight states, was formed by collapse during the eruption of about 50 cubic kilometers of ash and pumice only 6,900 years ago. The volcanologist Howel Williams clarified some of the concepts of caldera formation in his milestone 1942 study of Crater Lake. The 8.9-kilometer-diameter crater, perched at the summit of a young volcanic field, is partly filled with a remarkably clear, 590-meter-deep, intensely blue lake. Exposed in the caldera wall are a variety of small stratovolcano cones and lava flows. After the caldera formed, an eruption or eruptions within the crater formed Wizard Island, a cinder cone that rises above lake level. A 53-kilometer (33-mile) road circumnavigates the caldera, with many interpretative stops. Numerous trails around and into the caldera make Crater Lake excellent for viewing the products of eruptions.

Getting there: Driving north from California, there is a choice of two routes to Crater Lake: (1) Take U.S. highway 97 north from Klamath Falls for 37 kilometers (23 miles); turn left on Highway 62 and continue northwest for 55 kilometers (34 miles) to the park headquarters; (2) Leaving Interstate 5, turn north on Highway 62 at Medford and follow it for 95 kilometers (59 miles) to the park headquarters. From Eugene, follow Highway 58 to Highway 97 south, turning west at Crater-Diamond Lake Junction toward the park's north entrance.

Note well: If you want to see all the features of Crater Lake, go during the summer; the snow is very deep in the winter. Only a few facilities are kept open during winter, at which time the park is accessible from the south entrance only.

Mount Shasta, California

Traveling along Interstate Highway 5 from California toward Oregon on a clear day, Mount Shasta, at an elevation of 4,315 meters (14,161 feet) dominates the view on the northern horizon. The freeway touches the western base of this impressive stratovolcano at the town of Mount Shasta and cuts into the base of Black Butte, a dacite dome peripheral to the volcano. Mount

Shasta consists of four large overlapping cones constructed during the last one hundred thousand years, the last historic activity being in 1786. Most of the activity has consisted of eruptions of andesite lavas and the construction of thick domes and flows. The mountain can be climbed by rugged hikers and viewed from all sides from paved highways. Sometimes outnumbering the hikers are the cult members who believe the mountain is sacred or magical and that an advanced culture of "Lemurians" live beneath the mountain.

Medicine Lake Highland / Lava Beds National Monument, California

Devotees of volcanoes and birds need to be in northern California during the fall season. The trip south from Klamath Falls, Oregon, toward the town of Tulelake, California, crosses the spectacular Tule Lake National Wildlife Refuge, which is situated in a fault block basin peppered with basaltic volcanoes. The basin is home to over a million birds, mostly waterfowl, during the fall migration. Within the former lake basin there are tuff rings, and on the southern edge of the wildlife refuge is Lava Beds National Monument, with young cinder cones, abundant lava flows, and three hundred lava tubes. The monument is on the north flank of the Medicine Lake Highland, a broad basaltic shield with a caldera at its summit and several large rhyolite flows (Glass Mountain, Medicine Lake Glass Flow, Crater Glass Flow, and Little Glass Mountain). The ground surface is littered with pumice from eruptions that occurred about eleven hundred years ago.

Getting there: The Tule Lake basin and Lava Beds National Monument are easily reached from Klamath Falls on Highway 39. Watch for signs to the National Wildlife Refuge and to the National Monument. The Medicine Lake Highland is reached only by gravel forest roads and should be approached only after purchasing a Forest Service road map.

Note well: The nearest services are in Tulelake, California. Be prepared to camp if you stay in the area.

Lassen Peak / Lassen Volcanic National Park, California

The last volcanic eruption in the contiguous United States prior to Mount St. Helens was at Lassen Peak in 1914. Lassen Peak and the hummocky ridge north of it (Chaos Crags) are large dacite domes located on a high volcanic plateau. Pyroclastic flow, avalanche, and lahar deposits are associated with these domes. Geothermal areas, with boiling springs and mudpots, are located immediately south of the peak. Much of the park to the east of Lassen Peak is mantled with cinder cones and basaltic lava flows and is an excellent region for nonstrenuous hiking. Good interpretative exhibits are located throughout the park, and there is a trail to the summit of Lassen Peak.

Getting there: The park can be reached from Redding by Highway 44 (north entrance) or from Red Bluff or Susanville on Highway 36 (south entrance). Because of heavy snowfall, the north-south road across the park is open only from midsummer to early fall.

Note well: If you hike to the summit of Lassen Peak, be prepared for rapid weather changes. Around the geothermal areas, stay on marked trails, and be careful around springs and mudpots; a mistake can be fatal.

Long Valley Caldera / Mono Craters / Mammoth Peak / Inyo Craters, California

Some of the most beautiful drives in California are along the base of the eastern escarpment of the Sierra Nevada. One of these also coincides with the largest volcano in California—the Long Valley Caldera. Long Valley Caldera was formed during the eruption and emplacement of the Bishop Tuff, about seven hundred thousand years ago. Highway 395, between Bishop and Lee Vining (where a road goes west into Yosemite National Park), bisects this 17-by-32 kilometer (10.2-by-19.2 mile) caldera, which has been the site of uplift and seismic activity during the last ten years. Crossing the western side of the caldera are the Inyo Craters and rhyolitic domes (last activity about six hundred years ago). Immediately north of the caldera (also along the highway) is the 12-kilometer-long (7.2 miles) chain of rhyolite domes called Mono Craters, which overlie widespread pumice and ash deposits. Along the southwestern rim of the caldera is Mammoth Mountain, a young stratovolcano and site of one of the largest ski areas in the United States. Recently, carbon dioxide build-up around Mammoth Mountain has killed patches of trees, and there have been reports of the gas accumulating in buildings. Consult the Forest Service office on conditions of camping and hiking on Mammoth Mountain. The office is located just east of the town of Mammoth.

Getting there: Take U.S. Highway 395 either north from Los Angeles or south from Reno, Nevada. Either approach has spectacular geology and scenery.

Yellowstone National Park, Wyoming, Idaho, Montana

Yellowstone National Park is a major tourist destination for good reason. The very large calderas in the park have erupted about 6,000 cubic kilometers of material over the last 2.5 million years and have left a still-hot magma body beneath the surface, which provides spectacular geothermal displays at the surface. Although there are three overlapping calderas, they are difficult to recognize because of their size. But within their confines, it is easy to see the rhyolitic lava flows and dozens of thermal areas (boiling springs, geysers, travertine deposits, etc.). The volcanoes are clearly described in the museums, at the many roadside markers, and in the publications available within the park.

Getting there: Yellowstone can be reached via Highways 16 (from Cody, Wyoming), 212 (from Red Lodge, Montana), 287/89 (from Jackson, Wyoming), 89 (from Livingston, Montana), and 20/191 (from Idaho Falls, Idaho). There are numerous roads within the park.

Note well: Stay on marked trails within the thermal areas; you could be boiled! Do not feed the wild animals, especially the bears.

Hawaiian Islands / Hawaii Volcanoes National Park

All soil and rock on the Hawaiian Islands is derived from volcanic eruptions except the coral reefs that fringe the islands. For the traveler to the island of Hawaii, it is recommended that you buy one or several of the guides listed in the references for this section and stop at every museum and interpretative center. Since the activity and the shape of the land are constantly changing, use the information phone number listed for the U.S. National Park Service or follow the daily status reports posted in museums. Of particular interest are the excellent museums at Hawaii Volcanoes National Park headquarters and at the Hawaiian Volcano Observatory.

Note well: Weather at higher elevations can be cold and wet, regardless of the island's tropical setting. Follow directions listed in the introduction for visiting lava flows.

Valles Caldera / Bandelier National Monument, New Mexico

A 22-kilometer-diameter (14 miles) drive-through caldera, pyroclastic deposits, youthful rhyolite lava domes and flows, and a large, structurally resurgent dome consisting of pushed-up caldera floor characterize the Valles Caldera of northern New Mexico. Most tourists focus upon the archaeological wonders of Bandelier National Monument, located near Santa Fe. They do not know about the caldera but about the caves carved by the Pueblo Indians, who were attracted to the volcanic blessing of easily excavated ignimbrites deposited during the eruption that formed the collapsed Valles Caldera one million years ago. Tourists can add a major caldera and associated deposits to their anthropologically oriented trip with advanced planning.

Getting there: Go north on Highway 285/84 from Santa Fe, then turn west on Highway 502/4; follow the signs to the monument. Continuing west on Highway 4 to Jemez Springs will take you through the caldera and the surrounding Jemez Volcanic Field.

References for Canada and the United States

Alt, David D., and Donald W. Hyndman. *Roadside Geology of Northern California*. Missoula, Mont.: Mountain Press Publishing Co., 1975.

Corcoran, Thom. *Mount St. Helens—The Story behind the Scenery*, Las Vegas, Nev.: K. C. Publications, 1985.

Decker, R., and B. Decker. *Volcano Watching*. Hawaii: Hawaii Natural History Association, 1980.

————. *Road Guide to Crater Lake National Park*. Mariposa, Calif.: Double Decker Press, 1988.

Fritz, William J. *Roadside Geology of the Yellowstone Country*. Missoula, Mont.: Mountain Press Publishing, 1985.

Harris, Stephen L. *Fire and Ice: The Cascade Volcanoes*. Seattle: The Mountaineers, 1980.

Pringle, Patrick T. *Roadside Geology of Mount St. Helens National Volcanic Monument and Vicinity*. Washington Department of Natural Resources Circular 88. Olympia, Wash., 1993.

U.S. National Park Service, U.S. Department of the Interior, Hawaii Volcanoes National Park. Address: Superintendent, Hawaii Volcanoes National Park, HI 96718. You can receive eruption information by calling (808) 967–7977.

U.S. National Park Service, U.S. Department of the Interior, Lassen Volcanic National Park, California. Address: Superintendent, Lassen Volcanic National Park, P.O. Box 100, Mineral, CA 96063–0100, (916) 595–4444. Three fact sheets as well as a park newspaper (*Park Guide*) are available upon request.

Wood, C. A., and J. Kienle. *Volcanoes of North America: United States and Canada*. New York: Cambridge University Press, 1990.

Mexico and Central America

Mexico and Central America have thousands of volcanoes. The few mentioned here are based on the volcano's fame, its accessibility, and the presence of tourist facilities.

Parícutin, Mexico

Parícutin Volcano was born in February 1943 in the state of Michoacán. It was named Parícutin after a nearby village that it destroyed. The eruptive activity lasted nine years, producing a cinder cone 424 meters (1391 feet) high and lava flows that covered 25 square kilometers (9.7 square miles). Two villages were buried and their residents displaced. Many aspects of the eruption were documented, from the physical processes to its cultural impact on the Tarascan Indians who fled from the lava flows and ash fallout (the subject of an excellent book edited by Luhr and Simkin in 1993, the fiftieth anniversary of the eruption). The volcano has spawned a tourist industry, and you can visit the volcano and buried villages with local guides.

Getting there: The nearest town with facilities for tourists is Uruapan, slightly off the direct route between the cities of Guadalajara (300 kilometers; 186.3 miles) and Morelia (135 kilometers; 83.8 miles). Parícutin is reached by traveling north along Highway 37 for 16 kilometers (10 miles), then turning left on a side road (for 19 kilometers; 11.8 miles) to the village of Angahuan. Tours can be arranged there or in Uruapan.

Note well: Leave belongings in secure hotels and automobiles in protected parking lots. Be sure to take bottled water or other bottled drinks with you wherever you go.

Popocatépetl and Iztaccihautl, Mexico

These composite volcanoes are about 80 kilometers (50 miles) southeast of Mexico City. With elevations of 5,452 meters (17,886 feet) (Popocatépetl) and 5,286 meters (17,342 feet) (Iztaccihuatl), both volcanoes are physical challenges. Driving to the lodges in the saddle between the two mountains, at an elevation of 3,950 meters (10,835 feet), will provide good views. At the time of this writing, Popocatépetl was erupting and access to the saddle was closed. Check with the Mexican Tourist Organization before planning a visit.

Getting there: Take Highway 190-D southeast from Mexico City (the highway to Puebla). Leave 190-D and go south on Highway 115 to Amecameca (the total distance from Mexico City to Amecameca is 65 kilometers; 40 miles). One kilometer south of Amecameca is a road east to Popo-Izta National Park; this road continues for 24 kilometers (15 miles) to Paso de Cortes. From Paso de Cortes, 2.6 kilometers (1.6 miles) of paved road leads to two lodges.

Note well: Leave belongings in secure hotels and automobiles in protected parking lots. Take bottled water on any hikes. You must be acclimatized to the high elevations before attempting to follow trails to the top of either volcano.

El Pico de Orizaba (Citlaltepetl), Mexico

Orizaba is Mexico's highest mountain with a summit elevation of 5,700 meters (18,700 feet). This enormous inactive composite cone is a popular destination for mountain climbers and dominates the horizons of both Puebla and Veracruz. As is the case for Popo and Izta, unless you are a serious mountaineer, you must be satisfied with distant views or a trip to villages at the base of the volcano.

Getting there: Take Highway 150-D east from Puebla toward Orizaba (a worthwhile tourist goal itself). For a closer look, turn north on Highway 140 to Acatzingo (the road to Jalapa); 34 kilometers (21 miles) north of Acatzingo is the turnoff to Tlachichuca, where a 21-kilometer-long (13 miles) rough gravel road goes to Piedra Grande (four-wheel drive recommended). Piedra

Grande, at an elevation of 4,260 meters (13,977 feet), is the starting point for mountaineers.

Note well: Leave belongings in secure hotels and automobiles in protected parking lots. Take precautions with water and be acclimated to and prepared for high altitudes.

Amatitlan and Atitlan Calderas, Guatemala

Amatitlan and Atitlan are enormous calderas located in the lush Guatemalan highlands. The calderas were formed by voluminous eruptions of pumice fallout layers and pyroclastic flows, which mantle much of the central highlands. Both have large easily accessible lakes reached by paved roads.

Amatitlan, a complex of calderas 14 kilometers by 16 kilometers (8.7 miles by 10 miles), is only 27 kilometers (16.8 miles) south of Guatemala City. The youngest of the overlapping Atitlan calderas formed about 84,000 years ago during an eruption of about 250 cubic kilometers (60 cubic miles) of pumice and ash. The views are picturesque and the hot springs pleasant. However, the lake is too polluted for swimming.

The 20-kilometer-diameter (12 miles) Atitlan Caldera is located 147 kilometers (91.3 miles) west of Guatemala City and contains one of the earth's most beautiful lakes. It is easy to locate, and is one of Guatemala's main tourist attractions. Swimming is permissible. The composite cones of Toliman, Atitlan, and San Pedro are located on the southern shore, and the colorful villages on their slopes are accessible by ferry or tourist boat.

Getting there: The highway to Escuintla and Puerto San Jose (CA-9) passes by Lake Amatitlan only 27 kilometers (16.8 miles) south of Guatemala City. The same highway leads to Volcan Pacaya, an active Strombolian cone on the southern rim of the caldera.

To reach Lake Atitlan, follow the Pan American Highway (CA-1) west from Guatemala City to Los Encuentros junction (117 kilometers; 72.7 miles); take the highway south to Solola and on to Panajachel (total of 17 kilometers; 10.6 miles), where most hotels and restaurants are located.

Note well: Do not travel alone; armed robbery is possible at Atitlan.

Composite Cones and Domes of Guatemala

Two persistently active volcanoes and one that is intermittently active are attractions to nature travelers within Guatemala. Just outside the western city of Quetzaltenango is the dacite dome complex of Santiaguito, which has been erupting since the 1920s from the side of the Santa Maria composite cone, which was partly destroyed by a catastrophic eruption in 1902. Activity consists of slow growth of the thick, massive lava flows, occasional explosive bursts from the crater, and small pyroclastic flows that burst from the flow fronts.

Pacaya is a cone located on the southern rim of the Amatitlan Caldera and is visible from parts of Guatemala City. It has been erupting persistently from 1961 to the present; activity ranges from Strombolian bursts and lava fountaining to explosive hydrovolcanic activity that follows heavy rainstorms. It can be climbed on a day trip (tour guide recommended).

The skyline near Antigua and Guatemala City is dominated by three volcanoes—Agua, Fuego, and Acatenango. The summits of all three are above 3,700 meters (12,139 feet). You need to be in good shape to climb them; bring camping gear and a delight in slogging up seemingly endless slopes. The rewards are the indescribable views from their tops. Fuego erupts explosively every few decades, and Acatenango had a phreatic eruption in the 1970s.

Getting there: Fuego, Agua, and Acatenango can be approached from Antigua. Inquire at the tourist information office in the city. The Santiaguito/Santa Maria trailheads are immediately south of Quetzaltenango.

Note well: Bandits can be found near the trailhead for Pacaya; the trip is worthwhile to see an active volcano, but go with a tour or a group. Do not travel if the forecast is for bad weather.

Composite Volcanoes of Costa Rica

There are active volcanoes within one day's drive from San José, the capital of Costa Rica. Most are visible and accessible during the dry season, December through April, but are usually wreathed in impenetrable clouds during the remainder of the year. The most accessible volcano is Poas, with continual fuming and occasional eruptions within its 3-kilometer (1.9 miles) diameter crater. There have been sporadic eruptions over the last forty years, including eruptions of molten sulfur. The volcano is closely monitored and access is closed when activity is imminent.

Arenal Volcano has been erupting intermittently for decades, with explosive activity, pyroclastic flows, and lava flows. There is an observatory, complete with a lodge, where eruptions can be watched in comfort.

Irazu Volcano last erupted in 1963, covering San José and surrounding area with ash. Although the volcano is not in eruption, the views are worth a trip to the summit. Roads wind upward through the clouds, passing alpine chalets and dairy farms on the way to the summit crater, which is at an elevation of 3,400 meters (11,155 feet).

Getting there: To reach Poas Volcano, travel north from San José through Heredia and Barva along the main highway to the Parque Nacional Volcan Poas. Signs are rare and highway markers nonexistent—ask directions as you go. From San José to Arenal Volcano is a day's drive and is best reached (on highways) by traveling northwest through Alejuela, Zarcero, Ciudad Quesada, and on to Fortuna de San Carlos, the town nearest the volcano. From Fortuna de San Carlos, the road winds around the north side of the volcano. A gravel road turns south near Lake Arenal toward the observatory and lodge. There is a river to ford and some steep climbs. The summit of Irazu Volcano is reached via an excellent paved road that proceeds north from the city of Cartago (follow signs to the Parque Nacional Volcan Irazu).

Note well: These volcanoes are all high, and weather can be cold and wet.

References for Central America and Mexico

Alvarado Induni, Guillermo. *Los Volcanes de Costa Rica.* Editorial Universidad Estatal A Distancia, San Jose, 1989. In Spanish. A well-done book on every aspect of each of Costa Rica's volcanoes; each section has a history of activity, maps, chemistry, and even legends.

Luhr, J. F., and T. Simkin, eds. *Parícutin: The Volcano Born in a Mexican Cornfield.* Phoenix, Ariz.: Geoscience Press, 1993. A truly delightful book about every aspect of this famous Mexican volcano.

Prahl Redondo, Carlos E., and Miguel Suarez Flores. *Guia de Los Volcanes de Guatemala.* Club Andino Guatemalteco, Guatemala, Central America, 1989. In Spanish. Includes maps, routes, estimated times of climbs, and background.

Secor, R. J. *Mexico's Volcanoes: A Climbing Guide.* Seattle: The Mountaineers, 1981. Prepared mostly for mountaineers, with routes, potential problems, and general background.

Yarza, Esperanza de De la Torre. *Volcanoes de Mexico.* Toluca, Mex.: Universidad Autonoma del Estado de Mexico, 1984. In Spanish. Information about Mexican volcanoes is preceded by an introduction to volcanoes and igneous rocks. Good background, but little information on trails or roads to the volcanoes.

South America

There are thousands of volcanoes along the Andean chain, from Colombia to southern Chile. The volcanoes include some of the highest mountains in the world, discouraging all but the most avid mountaineers, although on some of the volcanoes you can drive to 6,000 meters in elevation.

The most frequently visited South American volcanoes are those of the Galapagos Islands, where the earliest (and some of the best) volcanological studies were done by Charles Darwin. The youth of these basaltic shield volcanoes allows the visitor to see well-preserved calderas, ash deposits, tuff rings, and lava flows. Visitors occasionally may witness an eruption.

References for South America

Gonzalez-Ferran, O. *Volcanes de Chile.* Santiago, Chile: Instituto Geografico Militar, 1995.

Hall, M. *El Volcanismo en El Ecuador*. Quito: Biblioteca Ecuador, 1977.
de Silva, S. L., and P. Francis. *Volcanoes of the Central Andes*. Berlin: Springer-Verlag, 1991.

New Zealand

The North Island of New Zealand is bisected by the Taupo Volcanic Zone. Within this volcanic zone, there is high heat flow, geothermal areas, and many volcanoes. Outside the Taupo Volcanic Zone, there are young volcanic fields encompassing Auckland, north of Auckland (around Ngawha), Taranaki (Mt. Egmont) Volcano, and older volcanic fields on the South Island. The few thumbnail descriptions here concern the most frequently visited volcanic and geothermal areas of the North Island, beginning in the north in the Bay of Plenty and then moving south.

White Island

White Island, 50 kilometers offshore, is one of the northernmost volcanoes of the Taupo Volcanic Zone. The summit craters are the focus of intensive research on this nearly continuously erupting volcano. Fresh craters are formed every few years and the fumaroles are hot enough to soften drilling pipe. There was a sulfur mine in the crater that was destroyed by an eruption in 1914. There is a variety of volcanic and geothermal processes at work, but you must be extremely careful.

Getting there: You can reach the island from Whakatane by boat cruises during the day, or by helicopter.

Note well: This volcano is extremely dangerous and should be visited only with a reputable volcanologist as a guide.

Rotorua Area

Signs of volcanism are everywhere in the Rotorua area. Widespread volcanic ash beds, rhyolitic lavas, and large calderas are only part of the catalog of volcanic phenomena. The town of Rotorua has developed around a large geothermal area, which is an integral part of the community. Recommended sites include Mount Tarawera, famous for its brief but spectacular eruption in 1886; the Waimangu thermal area, which developed after the Tarawera eruption; Lake Tarawera and the buried village (Te Wairoa); and the Waiotapu thermal area. All of these sites have excellent displays and publications for the volcano tourist.

Getting there: Rotorua is 240 kilometers southeast of Auckland, via excellent highways. It is only 88 kilometers south of the coastal resort town of Tauranga.

Note well: In the geothermal areas, stay on the trails. Boiling springs are unforgiving if you fall in.

Taupo-Wairakei Area

Although as varied as Rotorua, the Taupo-Wairakei area has two types of geological attractions—one obvious, and one not so obvious. The Wairakei and Ohaaki geothermal power stations are very prominent, with their steam pipes, cooling towers, and steam plumes easily spotted from the highway between Rotorua and Taupo. Less evident is the fact that Lake Taupo is a caldera covering an area of 600 square kilometers. Lake Taupo Caldera was formed during an eruption only eighteen hundred years ago, which left a large area of the North Island covered with a thick layer of fine-grained volcanic ash. Start your tour of this area with a visit to the volcanic museum at the Wairakei Research Centre, located along the main highway, about 6 kilometers north of Taupo. Another good museum is located a few kilometers farther at the Wairakei Power Station visitor center. A large swimming complex at the AC Thermal Baths, located in Taupo, is a facility for activities ranging from professional health training to just relaxing in a hot pool.

Getting there: Taupo and Wairakei are 80 kilometers south of Rotorua and 280 kilometers southeast of Auckland.

Tongariro National Park

Some of the most spectacular and youngest composite cones in New Zealand are within Tongariro National Park at the southern end of the Taupo Volcanic Zone. Ngauruhoe, Tongariro, and Ruapehu volcanoes dominate the skyline, with elevations ranging from 1,968 to 2,797 meters. Historic eruptions of these volcanoes have produced ash fallout, pyroclastic flows, and lahars. Lahars have been the greatest hazard—a lahar and flood from the 1953 eruption of Ruapehu carried away a railway bridge and train, killing 153 people.

There is much to see and do in Tongariro National Park, but first purchase a guidebook locally. If you tire of volcano exploring, there is winter skiing on Ruapehu.

Getting there: Excellent roads lead to the base of or onto the volcanoes in the park. It is 370 kilometers south of Auckland and 270 kilometers north of Wellington.

Note well: At these higher elevations, the weather can be foul at times; be prepared with warm clothing and raingear.

References for New Zealand

Ballance, P. F., and I.E.M. Smith. *Walks through Auckland's Geological Past.* Geological Society of New Zealand, Guidebook no. 5, 1982.

Cox, G. J. *Fountains of Fire—The Story of Auckland's Volcanoes*. Auckland, New Zealand: William Collins, 1988.

Houghton, B. F. *Geyserland: A Guide to the Volcanoes and Geothermal Areas of Rotorua*. Geological Society of New Zealand, Guidebook no. 4, 1982.

Institute of Geological and Nuclear Sciences. Information Series. These are well-prepared color foldouts covering the volcanology of specific areas and are sold locally. An example is the brochure *Taupo the Volcano*, which is about New Zealand's largest caldera.

Indonesia

The Indonesian arc is one of the most active volcanic arcs in the world. It gave us two of the world's largest historic eruptions—Tambora in 1815 and Krakatau in 1883. The Indonesian volcanoes are highly explosive and rarely produce lava flows. Located along a plate margin, this widespread nation of about thirteen thousand islands has 170 active or dormant volcanoes and the earth's largest caldera, Toba, on the island of Sumatra, which erupted 74,000 years ago. The rich soil and tropical climate have supported the growth of a dense population and its cultural coexistence with active volcanoes. Nearly every city has a nearby, sometimes active, volcano. To name a few, Bandung, where the Volcanogical Survey of Indonesia headquarters resides, has Tangkuban Perahu beyond the city's northern limits; Yogyakarta has the frequently active Merapi Volcano, 25 kilometers to its north; Bali has Gunung Agung and Gunung Batur; and the latest tourist destination, Lombok, has Gunung Rinjani. Nearly all of these volcanoes can be reached by local bus or by package tours. Even Krakatau can be reached by boat from nearby beach resorts on the west coast of Java, but access has been limited since a tourist was killed last year on Anak Krakatau, the now active volcano growing within the caldera of Krakatau.

Galunggung

Galunggung, slightly less than 2,170 meters high, has an amphitheater-shaped crater open to the southeast. At the foot of the volcano, below the open amphitheater, is hummocky ground composed of about four thousand hummocks named the Ten Thousand Hills of Tasik Malaja. Prior to the eruption of Mount St. Helens, many thought that this strange landscape was formed by lahars, but it is now known to be the deposit of a large debris avalanche.

Gunung Kelud

The Javanese volcano, Gunung Kelud, has spawned many killer lahars within historic times. At one time, the summit of the volcano was a crater lake. Eruptions of pyroclastic debris through the lake mixed with the water and created destructive, hot lahars in 1811, 1826, 1835, 1848, 1864, 1901, and 1919. A tunnel was drilled through the flanks of the cone to drain the water from the crater and has had some success in lahar prevention, but it was extremely expensive.

Philippines

Nature has both blessed and cursed the people in the lively, vibrant Philippines. A tropical climate and reasonably good soil has led to the growth of a large population for this island nation, but a major strike-slip fault and twenty-one active volcanoes are the basis for substantial geologic hazards. Awesome volcanoes can be found from central Luzon, near Manila, to the southernmost end of the island of Mindanao. The setting is ideal for geothermal development, and new steam wells continue to feed generating plants for the energy-hungry nation. There are few guidebooks, but there is continual mention of the volcanoes or thermal springs within tourist guidebooks. The drive northwest of Manila to Angeles City will reveal the effects of the 1991 eruption of Mount Pinatubo. The eruption of this volcano closed military bases and devastated large tracts of farmland in central Luzon; this devastation continues with lahars and floods that accompany monsoon rains. The volcano is still dangerous and inaccessible, but overflights can be arranged.

Taal Volcano (Lake Taal) is a 21-by-13 kilometer caldera located only 35 kilometers south of Manila. In the center of the caldera lake is Volcano Island, where frequent eruptions drive the island's inhabitants to shore; it has erupted thirty-three times since 1572. The hydrovolcanic eruption on Volcano Island in 1965 left 137 dead and 165 missing. Lake Taal fills a prehistoric, but young, caldera complex that is surrounded by thick pyroclastic flow deposits, which reach as far as the southern edge of Manila. Resorts on the caldera rim and along the lake give dramatic views of this volcanic landscape. The Philippine Institute of Volcanology and Seismology maintains an observatory and small museum along the northern shore at Talisay (also a place to take small boats to Volcano Island).

Reference for the Philippines

Hargrove, T. R. *The Mysteries of Taal: A Philippine Volcano and Lake, Her Sea Life, and Lost Towns.* Manila: Bookmark Publishing, 1991.

Japan

Japan encompasses three major islands and hundreds of small volcanic islands. The three large ones are Hokkaido on the north; Honshu, the main, central island; and Kyushu on the south, each with many volcanoes. The country is a chain of volcanic islands forming part of the Pacific Ring of Fire. There are about two hundred volcanoes in Japan that are one million years old or less, sixty of which are active or in eruption. Japanese volcanoes are distributed in two major zones known as the East and West Japan volcano belts. One belt is an extension of the Kurile Arc that extends from the north to Hokkaido and links with the Honshu Arc that joins with the Izu-Mariana Arc to the central part of Honshu. Volcanoes of these arcs are caused by subduction of the Pacific tectonic plate along the Kurile-Japan-Izu-Mariana Trenches. The East Japan volcanic belt extends along the Japan Sea west of Honshu and is linked to the Ryukyu Arc through the volcanoes of Kyushu. The Japanese, with an informed public and many first-rate volcanologists, have learned to live with their volcanoes. The active volcanoes near large population centers are hooked up to seismometers and videocam recorders and are under constant surveillance. Their characteristics are well known, therefore a change in behavior is instantly recorded, and the populace is warned of possible eruption and evacuation.

Many volcanoes are national parks, shrines, or playgrounds and are attractions for thousands of Japanese tourists each year. But unless the volcano traveler knows the Japanese language well, tours must be guided. Translating Japanese signs along the road to find your way, or even to find a hotel, can be difficult if you are driving without a guide.

Bandai Volcano

Bandai Volcano is a horseshoe-shaped volcano located about 230 kilometers north of Tokyo in beautiful wooded country. The volcano erupted on July 15, 1888. Like Mount Saint Helens, part of the cone collapsed northward in a gigantic debris avalanche and sent forth a pyroclastic blast. The avalanche divided into two lobes that spread up and down Nagase Valley, at the foot of the volcano, burying the valley, villages, and all living things beneath hummocks of the volcano. The blast uprooted large trees or snapped them off at ground level. Blockage of natural streams caused formation of four large lakes, forcing the evacuation of several villages that had survived the avalanche and blast. Four hundred and sixty-one people that lived in Nagase Valley were killed. Vegetation has since returned in the century since the eruption. The hummocks caused the formation of many small lakes. The knowledgeable volcano traveler who has visited Mount St. Helens will note many similarities. Much of this land has been set aside as a beautiful park that attracts many visitors each year.

Fuji Volcano

The most famous volcano in Japan is Mount Fuji (Fuji-san). It is part of a chain of volcanoes within the Fuji Volcanic Zone that extends south to form the Izu

Peninsula. In the last one thousand years, it has erupted at least thirteen times, the earliest being in 781 and the most recent in 1707. The most violent eruptions occurred in 800, 864, and 1707. Millions of people now live within reach of its possible hazards.

Mount Fuji, 100 kilometers southwest of Tokyo, is the highest volcano in Japan, rising to 3,776 meters in elevation, and is 50 kilometers broad at its base. At its top is a circular crater 500 meters across and 250 meters deep. Mount Fuji has a simple conical shape, but it is actually a complex of three superimposed cones. There are also more than one hundred parasite cones and flank openings.

It is a sacred mountain, and each year is visited by tens of thousands of pilgrims, who ritually ascend Mount Fuji with its ten stations. The flourishing cult of Mount Fuji, which has more than thirteen hundred shrines, has made the volcano Japan's symbol. For as long as the Japanese can remember, people have been climbing Mount Fuji and its graceful shape has inspired Japanese artists and poets for centuries.

Hakone Volcano

Hakone Volcano borders Mount Fuji to the east and is north of many volcanoes that form the Izu Peninsula. It is a beautiful complex volcano designated as a national park in 1933. The volcano has many beautiful sights and hot springs with comfortable hotels, excellent paved roads, and railroads to the region. The volcano traveler can reach Hakone though three major entrances by car. The east entrance is a winding road through deep canyons. The west entrance is across a flat slope of the volcano and offers breathtaking views of the Pacific Ocean. The northern entrance gives the perfect conical view of the volcano.

Hakone is composed of two overlapping calderas and seven post-caldera cones. The main body of the volcanic edifice is about 20 kilometers long and 15 kilometers wide. There is a beautiful crater lake and abundant volcanic features.

More than forty geothermal areas are known in the Hakone and Izu district that extends 80 kilometers by 30 kilometers. All of the geothermal areas have hot-spring resorts.

Izu-Oshima Volcano

Oshima Volcano, an island in Sagami Bay, is 100 kilometers southwest of Tokyo. It is a popular vacation island frequented by Japanese tourists and is easily accessible by a forty-minute plane flight. Boat trips are scheduled daily from several harbors of Honshu, including Tokyo, and take several hours. Some are all-night trips. Taxis, buses, and rent-a-cars are available to visit the volcanic features on the island. Oshima is the biggest island of the seven Isu islands. The island has been occupied by ancestral Japanese for as long as nine thousand years.

Oshima has a shield-like shape with an area of about 92 square kilometers. At its summit is a collapsed caldera that measures 3 by 4 kilometers.

Sakurajima

If the volcano travelers want to witness small eruptions that puff from the summit of a volcano, they should travel to the peninsula of Sakurajima located at the northwestern end of Kagoshima Bay on Kyushu. At times, explosions can be heard. It is possible to drive to its foot and walk on paved pathways to observation points near the volcano. Sakurajima is very active and explodes every few hours, a healthy escape valve. Without its escape valve, it can be violent, the last cataclysmic eruption being in 1914, when several island villages were destroyed.

References for Japan

Aramaki, S. *Field Excursion Guide to Fuji, Asama, Kusatsu-Shirane, and Nantai Volcanoes.* Symposium on Arc Volcanism. Tokyo and Hakone, Japan: Volcanological Society of Japan, 1981.

Oki, Y., S. Aramaki, K. Nakamura, and K. Hakamata. *Volcanoes of Hakone, Izu, and Oshima.* Hakone, Japan: Hakone Town Office, 1981.

Italy

Between Italy's Tyrrhenian coast and the Apennines, running from Florence to Naples, is a chain of very young volcanoes; most are large, low-rimmed calderas, some containing lakes. The world's first geothermal generating plant, at Larderello, located only 35 kilometers west of Siena, is at the northern end of this chain. There are no volcanoes at the surface here, but there is a large, rather shallow magma body under the geothermal field. Among the better-known volcanic fields are the Vulsini complex, the Vico Field, the Sabatini Field, and the Alban Hills located on the southern edge of Rome. Many of the famous hill towns of Italy, such as Orvietto, are located on the cliffs and mesas of pyroclastic flow deposits. Some of the better-known wines are from grapes grown on volcanic soils. The Pope's summer palace at Roca di Papa is on the rim of a young caldera in the Alban Hills, overlooking a caldera lake. Most of the basic building stone in Rome and Naples is ignimbrite. It cannot be polished and is often an ugly yellowish-brown color, but the ignimbrite occurs in such local abundance that it is used for the basic building block and is then faced by plaster or polished marble or travertine.

Traveling south to Naples and the famous Mount Vesuvius, the highway passes along the edge of Roccamonfina, yet another large and young caldera, where there are many rustic villages nestled on its slopes and in its crater.

There are few guidebooks available in the bookstores for Italian volcanoes except the one listed in the references. However, local well-illustrated booklets are usually available on-site at each volcano.

Volcanoes of the Bay of Naples

The vibrant, historically exciting cities lining the Bay of Naples aren't just *near* volcanoes; they are *on* them and very likely could not exist without them. Amid the tumult of automobiles and motorbikes typical of Naples and the other cities that encircle the Bay of Naples, there are many types of volcanoes and volcanic deposits.

Vesuvius and the famous Roman ruins at Pompeii, Herculaneum, and Oplontis caused by the volcano's A.D. 79 eruption are the most popular with tourists and can be visited by train and bus or by car. A short hike to the summit rewards the visitor with an impressive crater and a remarkable view of this part of Italy. On the way down the mountain, the visitor should stop at the Vesuvius Observatory.

Starting at the western edge of Naples are the overlapping 35,000-year-old Campanian and 11,000-year-old Yellow Tuff Calderas, now filled with the numerous tuff rings of the Phlegrean Fields (the youngest being Monte Nuovo, which erupted in A.D. 1568); geothermal areas such as Solfatara Crater and Mofete; and Lago Averno, the harbor for the Roman fleet in Pliny's time and considered by some to be the entrance to hell. The Phlegrean Fields are also rich in history, from the Cave of the Sybil, a Bronze Age oracle, to the Serapeum, a sea-level Roman marketplace. The central part of the caldera on which Pozzuoli is built rises and sinks with the rising and sinking of magma beneath the caldera like some breathing monster. It is not easy to travel within the Phlegrean Fields because of the dense population that lives within and along the rims of these young volcanoes, but patience will be rewarded with excellent volcanic sites.

Take one of the many short ferry trips from the Naples harbor to the volcanic islands of Procida or Ischia. The small island of Procida is made up of overlapping tuff rings and can be explored on foot. Ischia is a larger, more complex volcanic island and comprises the 788-meter-high Monte Epomeo, which was formed by uplift along fault planes; lines of craters and lava domes; and abundant thermal spas. It can be explored by public bus, by rented car, or by foot.

Mount Etna

Mount Etna, with an elevation of 3,340 meters, is Europe's largest active volcano. It dominates the geography and history of northeastern Sicily. When it erupts, it often plays havoc with ski facilities near its summit and occasionally with the towns and villages on its lower slopes. Most of the activity at the summit is Strombolian to Vulcanian, and the lower slopes are affected by lava flows. Mount Etna is definitely worth exploring by car. Start along its lower slopes at the seaside, then proceed toward the barren summit through orchards, gardens, and grazing land. The summit area is accessible by walking trails and by a ski lift from Rifugio de Sapienza. Because the terrain is constantly changing, guides will take you from the ski lift to the areas of most recent activity by four-wheel-drive vehicles.

Aeolian Islands

This chain of volcanic islands located north of Sicily is accessible only by ferries or hydrofoils (*aliscafi*) from Messina, Milazzo, or Reggio di Calabria. There are seven main volcanic islands, but only three have ample tourist facilities—Vulcano, Lipari, and Stromboli.

Vulcano is dominated by three features: Il Piano, a filled crater that forms a plateau; Vulcano, a hydrovolcanic tuff ring, which last erupted in 1890; and Vulcanello, a small cone on the northern tip of the island. The island is crowded with tourists in summer because of its good swimming and easy access by hydrofoil boats.

Lipari, immediately north of Vulcano, has a variety of pyroclastic deposits, many hectares of dune-shaped pyroclastic surge deposits, and silicic lava domes. It is the largest and most densely populated Aeolian Island. Ferries carry automobiles to Lipari.

Stromboli, the "lighthouse of the Mediterranean," has been erupting sporadically throughout historic time. There are no roads on the island. The ferry docks at the village of San Vicenzo, where there is accommodation and food. From there, a hike to the summit (924 meters) is a day trip. It can be cold at the summit, so take a windbreaker and sweater. The original "Strombolian" activity can be observed, but be extremely careful about locating your observation point. There are no railings or park rangers—like many places in Italy, you are on your own. An unforgettable experience is to spend the night at the summit, when showers of incandescent bombs light up the sky. If you go, take food and water and appropriate clothing. Make the ascent and descent during daylight hours; the trail can be treacherous.

References for Italy

Chester, D. K., A. M. Duncan, J. E. Guest, and C.R.J. Kilburn. *Mount Etna: The Anatomy of a Volcano*. Stanford, Calif.: Stanford University Press, 1985.

Gasparini, P., and S. Musella. *Un Viaggio al Vesuvio*. Napoli: Liguori Editore, 1992.

Greece

Forming a beautiful arc across the southern Aegean from the peninsula to the shores of southern Turkey are the volcanoes of Greece. The youngest include the peninsular Méthana, near Corinth, and the islands of Milos, Thera (Santorini), Nisiros, Kos, and Yali. This chain overlies the subduction zone where the European plate overrides the African plate. The last eruption was in the Thera volcanic field in 1950.

The most popular tourist destination in the Aegean is Thera (Santorini), where a volcanic field developed about one million years ago, but most of the visible deposits were emplaced during the last one hundred thousand years. Thera is actually three islands, once attached, that were separated by caldera collapses one hundred thousand years ago and during the seventeenth century B.C. These calderas were flooded by the sea and in their spectacular walls are perfect exposures of its volcanic history. One of Thera's main attractions is the Bronze Age town of Akrotiri, preserved beneath the deposits of the caldera-forming eruption that occurred in the seventeenth-century B.C. The Kamenis, islands in the middle of the caldera complex, were formed during eleven eruption episodes spanning the last two thousand years—they are often referred to by tour guides as "the volcano." They do not realize that the volcano comprises most of the islands of Thera.

You can reach Thera by ferry or jet aircraft. There are numerous tours available, but the local bus system is adequate.

Metric Conversion Table

Length
1 centimeter	0.3937 inch
1 meter	3.2808 feet
1 kilometer	0.6214 mile

Area
1 square meter	10.764 square feet
1 square meter	1.196 square yards
1 square kilometer	0.3861 square mile
1 square kilometer	247.1 acres
1 hectare	2.471 acres

Volume
1 cubic meter	35.314 square feet
1 cubic meter	264.17 gallons
1 cubic kilometer	0.240 cubic miles

Weight
1 gram	0.03527 ounce
1 kilogram	2.20462 pounds

Temperature
$(^\circ C \times {}^9/_5) + 32$	degrees Fahrenheit

About the Authors

Richard V. Fisher is Professor Emeritus of Geological Sciences at the University of California, Santa Barbara. Grant Heiken, a volcanologist at the Los Alamos National Laboratory, teaches courses in volcanology at the University of New Mexico. Jeffrey B. Hulen is a research geologist at the Energy and Geoscience Institute of the University of Utah, where he specializes in volcanic and other geothermal systems and associated gold mineralization.